TURKEY IN THE WORLD CAPITALIST SYSTEM

To my mother and to the
memory of my father

Turkey in the World Capitalist System

A Study of Industrialisation, Power and Class

Edited by
HUSEYIN RAMAZANOGLU
School of Social and Historical Studies
Portsmouth Polytechnic

Gower

Published by
Gower Publishing Company Limited,
Gower House, Croft Road, Aldershot, Hants. GU11 3HR,
England

and

Gower Publishing Company,
Old Post Road, Brookfield, Vermont 05036, U.S.A.

British Library Cataloguing in Publication Data

Turkey in the world capitalist system : a study
of industrialisation, power and class.
1. Capitalism——Turkey 2. Turkey——Economic
conditions——1960-
I. Ramazanoglu, Huseyin
330.12'2'09561 HC492

Library of Congress Cataloging-in-Publication Data
Main entry under title:

Turkey in the world capitalist system.

Includes index.
1. Capitalism--Turkey. 2. Industry and state--
Turkey. 3. Turkey--Economic policy. I. Ramazanoglu,
Huseyin, 1947- .
HC495.C3T87 1986 338.9561 85-17587

ISBN 0 566 05049 8

Printed in Great Britain by Paradigm Print, Gateshead, Tyne and Wear

Contents

Tables

Preface

This collection of essays has been put together as a contribution to the study of the development of capitalism on a world scale. The growing problems of reproducing this unevenly developed world system are found in every part of the world, and Turkey is just one arena for the emergence of the critical problems which the development of capitalism has generated. The study of Turkey's transformation from an inward-looking to an outward-looking economy forms the main focus of the book, but these articles should not be seen simply as a book on Turkey, rather this is an examination of one instance in the reproduction of capitalism on a world scale, and one which raises many general problems of how we understand the development of capitalism, and the operation of the world capitalist system.

I am grateful to Raymond Apthorpe, who read the whole manuscript in its initial and revised forms and encouraged the production of this book. I am also indebted to Dave Wilson and Cengiz Arin for comments and constructive criticisms. Feroz Ahmad made it possible to make editorial changes in his article; Caroline Ramazanoglu has engaged in critical discussion of all the arguments and issues in my contributions to this book, and has given most generously of her own working time, but I must comply with the tradition of claiming final responsibility for the ideas advocated in my articles, and for the editorial tasks undertaken in the production of this book. My thanks also go to the other contributors who have allowed me to use their work, and to the publishers who have kindly allowed the articles to be reprinted:

"Turkish Apertura", METU Studies in Development, Vol.7, Nos.3-4, 1980

"Short-Term Stabilisation Policies in a Developing Economy: The Turkish Experience in 1980 in Long-Term Perspective", METU Studies in Development, Vol.8, Nos.1-2, 1981

"Military Intervention and the Crisis in Turkey", MERIP Reports, No.93, January 1981

Southsea, 1985 Huseyin Ramazanoglu

Contributors

FEROZ AHMAD
Professor of History, University of Massachusetts, Boston

KUTLAY EBIRI
Formerly of the Turkish State Planning Office and the Department of Economics, Middle East Technical University, and currently with the World Bank

CAROLINE RAMAZANOGLU
Formerly of the universities of Keele, Brunel, and Middle East Technical University, and currently Senior Lecturer in Sociology, Goldsmiths' College, University of London

HUSEYIN RAMAZANOGLU
Formerly of the University of Istanbul, Middle East Technical University and the University of East Anglia, and currently Senior Lecturer, School of Social and Historical Studies, Portsmouth Polytechnic

FIKRET SENSES
Assistant Professor, Department of Economics, Middle East Technical University

Introduction

HUSEYIN RAMAZANOGLU

The development of Turkish capitalism has not often been considered to be of great interest to scholars of development studies in general, nor even to experts on the Moslem Middle East. Isolated by Ottoman history, language and culture from the West, and by Republican history and political choice from the East, Turkey may seem to stand as a unique case, of interest only to Turks and a narrow circle of Turkish-speaking scholars. For other scholars, Turkey is a typically underdeveloped area, of no particular interest as its unique character has been absorbed into a wider Third World or capitalist periphery.

These views, however, are seriously misleading, and in this book the development and upheavals of Turkish capitalism are presented both as of interest in their own right, and as of an essential part of the development of world capitalism. In order to understand the world we all live in, (or may very soon die in), we need to understand the world capitalist system as a whole, and the various parts which comprise it, of which Turkey is one.

Turkey is a rapidly industrialising country, but its incorporation into the world capitalist system, and its disadvantaged place in the international division of labour, has created economic, political and social problems which are common in other parts of the world. At the root of these problems lies the fact that the development that is taking place in the world today is very largely capitalist. However we would assert that the principle logic of capitalist development is that it is always uneven. This principle applies to the development of capitalism within the confines of a national economy, as well as to the development of capitalism on the world scale. There are some countries of the world which are obviously less developed than others; some areas are obviously highly developed and enjoy a dominant position in the world capitalist system. The fact remains, nevertheless, that there is uneven development all over the world, and the development of capitalism cannot be understood by ignoring this reality and denying the existence of capitalist development in all parts of the world. The schools of thought which have dominated development studies since the early 1970s, the various dependency and underdevelopment approaches, see the world as divided into "developed" and "underdeveloped" areas. In many respects Turkey may appear to be typically underdeveloped, but this appearance of underdevelopment is misleading and cannot be taken for granted. A systematic analysis of the development of its economic and political structures, makes it clear that while there are many similarities with other "typically underdeveloped" nations, there are also significant dissimilarities. This makes the use of the label "underdevelopment" problematic, since in many respects Turkey has not been subject to what are widely thought of as typical processes of "underdevelopment".

There has been a tendency in development studies for the dominant theories to be general theories and to emphasise the common characteristics of typical development processes. The case of Turkey is used in this book to show the inadequacies of world-wide generalisations of this sort, and to demonstrate that the questions which need to be asked in order to understand the development of Turkish capitalism, also need to be asked of other cases, if we are to arrive at a proper understanding of the world capitalist system.

There is an obvious temptation to generalise when evidence of apparently similar experiences and development processes is readily to hand, but there is a danger that such generalisations will prevent us from recognising the diversity of forms which development can actually take. Our understanding, not only of Turkish development, but also of development elsewhere, is likely to be much more productive if we avoid starting from the assumption that there are general forms of dependency relationships in which "developed" countries actively "underdevelop" other countries. Instead each case needs to be located in the uneven development of world capitalism as a system. Rather than imposing general labels on this development, we should aim to understand the development of world capitalism by trying to explain the varying forms which capitalism takes in different situations and at different times. Turkey is one part of the world capitalist system, and needs to be understood as a unique part of that system, yet as a part which is in a complex set of interrelations with other parts.

In this book, the focus is on the development of Turkish capitalism, within the the development of the world capitalist system. Turkey is taken to be a specific case of a social formation which is in the process of a rapid transformation. While all capitalist social formations must undergo a transformation in which capitalism emerges, transformation is not a uniform or entirely predictable process. The rise of capitalism in Turkey has occurred as a transformation which is unique to Turkey, but there are also many similarities with the rise of capitalism in other parts of the world. The upheavals experienced by Turkey since early 1970s are part of the world-wide problems of continuous capital accumulation and the consequent restructuring of the world economy that is taking place today. Turkey, undoubtedly, shares important features with many other Newly Industrialising Countries of the world. They are experiencing similar economic, political and social upheavals, and their people are suffering military rule, political repression, and the dismantling of "democracy". The Turkish experience in coping with such problems not only provides a highly relevant example for the analysis of capitalist development but also provides lessons from which other Newly Industrialising Country could profit from, despite the undemocratic nature of this example. The case of Turkey is, therefore, of general interest to those who wish to understand the development and operation of the world capitalist system, since both the similarities and the differences need to be explained. It is by careful analysis of the specific transformation processes which result in the rise of capitalism in particular social formations, that we will begin to achieve a theoretical understanding of the uneven development of world capitalism and its various consequences. The articles in this book are intended as a contribution towards such an understanding.

The authors whose articles are collected here do not share a common theoretical framework. Indeed, it is unlikely in such a contentious field that much unanimity could be found at present. What the authors do have in common is the aim of understanding the crisis of Turkish capitalism in the late 1970s and how it has come about. The articles address, in different ways, the themes of industrialisation, political power and social class, and the relationships between them. These are all aspects of the transformation process, and are scrutinised by the contributors from the perspective of somewhat different schools of thought, and different areas of expertise.

The essays which follow have been written with different aims. Some have been published before, while some have been specially written for this volume, but they have been brought together because in different ways they contribute to our knowledge of Turkey's problems as a constituent part of the world capitalist system. They all analyse aspects of the genesis of the crisis in Turkish capitalism, its current outcome, or probable future development. From their different perspectives they are all concerned with the problems generated by the need to shift Turkey from an inward-looking economy based on import-substitution, to an outward-looking economy based on export-promotion.

While the history of the need for this shift and the difficulties which it has created, may seem an internal Turkish problem, there are a number of other countries which are experiencing similar problems, The example of the way in which these problems have developed in Turkey should, therefore, be of interest not only to students of Turkish affairs, but also to those who are interested generally in the political economy of capitalist development, and in the analysis of the specificity of social formations within the totality of the world capitalist system.

Chapter 1, Current Problems in Conceptualising the Uneven Development of Capitalism, by Caroline Ramazanoglu and Huseyin Ramazanoglu, is intended to place the study of the development of Turkish capitalism in the general context of a critique of theories of capitalist development, the world capitalist system and the pre-requisites for its reproduction. Readers who are familiar with the debates on development and underdevelopment, but not with Turkey, should be able to see the relevance of the Turkish case. These theories are not dealt with individually in great detail, rather the basic assumptions on which each school depends are summarised, explained and criticised, and some suggestions are made on the need for an alternative approach to the explanation of capitalist development.

While there is general agreement among social scientists that a world capitalist system exists, there is very little agreement about what this system is and how it operates. We cannot understand Turkish capitalism in isolation from the capitalist system of which it is a part. In this chapter we consider some of the general theoretical issues which need to be resolved before going on to look in more detail at the case of Turkey. We review briefly the main theoretical views on the nature of the world system which have come from the modernisation and the dependency school and, more recently, from the contributors to the debate on the articulation of modes of production. As the dependency school has come increasingly under attack, the articulation of modes of production debate has become more elaborated, although many contributions remain obscure because of the abstracted level of the debate. We suggest that while the contributions of these schools have been fruitful in many ways, their basic premises are misconceived and unnecessarily limiting; in short, they are inadequate if we want to explain the development of Turkish capitalism in the context of the world capitalist system.

While the articulation debate has provided a trenchant critique of dependency theory (whose adherents would not recognise the existence of Turkish capitalism) the debate has arisen within a particular framework of assumptions which has resulted in an unwarranted emphasis being placed on an assumed general transition to capitalism. In spite of the verbal complexity with which they are usually presented, these assumptions can be summarised quite simply, and the authors show that these assumptions have prevented adequate attention being paid to the relation of general historical processes to specific historical situations, and the real struggles in which people are or are not actually engaged. In this chapter we reconsider the general problems of conceptualising the uneven development of capitalism on a world scale. We argue that if we want to understand the general historical processes of industrialisation, and the production of political power and social classes, we can only study them in specific historical situations. Industrialisation, is a characteristic of the world system, but it only occurs and, therefore, can only be studied, in particular social formations. This is equally true of the changing balance of political power, and the development of classes and class struggles. If we want to understand the world capitalist system, we can only do so by studying its various parts and regulating agencies. Turkey is just one part of this enormously complex and diverse system.

It must be emphasised that we do not argue that all analysis of the world system should be restricted to the piecemeal analysis of unique individual cases. Rather, the world system must be analysed on two interrelated levels. At a general level the logic and consequences of capitalist development must be the same, that is recognisably capitalist. At the same time, the actual processes of development give rise to different social formations which, while clearly dominated by the capitalist mode of production differ from each other, and over time, in quite specific ways. A theoretical understanding of the uneven development of capitalism must take into account the relationships between these general processes and actual historical forms. This point is not a new one, but it is our contention that the implications of taking this position have become lost in the confusions of competing development theories and that the relationships between general processes and historical forms remain problematic.

From this general theoretical discussion, we move to a brief examination of the operation of the world capitalist system and, especially, the role of the IMF as its chief regulating agency. We argue that Turkey, like any other social formation in the world capitalist system, has a position within the international division of labour. It is this division of labour which forms the basis of the world capitalist system, and determines its nature. Just as it would be a mistake to look at Turkey in isolation from the world capitalist system. So it would be a mistake to examine the world capitalist system as if it had a separate existence from the social formations of which it is comprised. The world capitalist system does exist as a system with interconnected parts, but it does not exist independently of these parts. When we talk of the world capitalist system, we are referring to the social, political and economic activities occurring in a multiplicity of social formations which differ from each other in a variety of ways. The world system is plagued by contradictions, but since the system is only the sum of its parts, these contradictions must be rooted in its parts.

The main characteristics of the world capitalist system, according to this view, are briefly reviewed. Its instability is explained. Since the system is neither unified nor harmonious in its operation, it has been important for the dominant powers in the system to define the needs of the system as a whole in ways which are consistent with their interests. This has led to the emergence of a number of international agencies which help to regulate the expansion of capitalism on a world scale. The emergence and nature of the main agencies are examined, and they are seen as political in essence and economic in form. The role and policies of the IMF are analysed in order to illustrate these points.

The IMF is the most important regulating agency for the international monetary system, and international money markets are greatly influenced by the stand taken by the IMF in relation to member states. The IMF has had a chequered history, however in recent years it has emerged as the leading international economic regulatory institution engaged in trying to maintain stability in the world system. The Fund's policies are analysed and the purpose and consequences of its interventions are discussed. We argue that all the measures taken by the IMF have political implications. It is seriously misleading to interpret them as guidelines for economic recovery based on value-free economic analysis. IMF policy packages are offered to states which are in difficulty on the assumption that they will continue to promote capitalist development and to fulfill their designated role in the maintenance of the world capitalist system.

The use of the term "world capitalist system" raises questions about the place of the socialist states. This issue is considered, together with a critical discussion of the Soviet model of development and its popularity in many "undeveloped" economies today.

This general theoretical examination of development within the present world capitalist system forms the basis for the overall framework of the book. Chapters 2-8 in this book all deal specifically with Turkey, but since the capitalist system can only be understood as a whole, Turkey has to be understood as a part of this whole. While Turkey,like all the other social formations which constitute the world system, makes its own distinctive contribution to the reproduction of world capitalism, it is simultaneously determined by the general level of contradictions which are generated in the world system. This immensely complex process obviously has to be understood if we want to understand how world capitalism works, yet our knowledge of this processes is relatively limited, and tends to be mystified by the use of universal generalisations.

Chapter 2, A Political Analysis of the Emergence of Turkish Capitalism, 1839-1950, by Huseyin Ramazanoglu, is an examination of the development of Turkish capitalism and the genesis of the crises of Turkish capitalism, which are examined in greater detail by other contributions to this book. It starts by looking at the significance of the Tanzimat and the Young Turk periods which determined the subsequent process of development of Turkish capitalism. and go on to examine the nature of Turkish capitalism under the new Turkish Republic, founded in 1923 by

Mustafa Kemal (Ataturk). This analysis of the nature of the Kemalist regime is crucial for an understanding of the conditions which led to the establishment and growth of Turkish capitalism, especially the etatist policies which influenced the basic direction of Turkish capitalism subsequent decades.

Tanzimat marks not merely a period of economic decline, but also a period in which the transfer of political power from the Ottomans to the ethnic minorities led to the final demise of the Empire after the First World War. During the Young Turk period, Moslem Turks outside the ruling circle began to view the loss of the Empire and its consequences with great concern. The concessions to foreigners and to the ethnic minorities made during the Tanzimat period created a nationalist reaction and led to moves by Moslems to regain economic and political power. Although the Young Turks did gain political power, in the long term their moves forced the Empire into the First World War in an attempt to break the Anglo-French stranglehold on the Empire. Defeat in the war allowed the Allies to occupy the Empire, triggering a popular struggle for independence which led directly to the establishment of the Republic of Turkey in 1923.

This chapter is mainly an analysis of the later period, up to 1950, in which Mustafa Kemal (later Ataturk) and his immediate associates and successors, managed to create new political, economic and ideological structures and practices, which swept away the remnants of the Empire and created the conditions which ensured the development of Turkish capitalism. In spite of pressures to keep the Turkish economy open to world markets, the Kemalist regime was successful in establishing a closed economy and a strategy of import-substitution. This was made possible by the Kemalist policy of etatism; a policy which encouraged direct state intervention in industrial development in order to create suitable conditions for the accumulation of private capital.

Although these policies were never easy to maintain, and etatism came under increasing attack during the late 1940s, being finally rejected by the Democrat Party after their electoral victory in 1950, the new Turkish state was founded and consolidated on the basis of etatist policies. Throughout this period, the dominance of etatism in the development of Turkish capitalism blocked the implementation of the alternative economic strategy of export-promotion in an outward-looking economy.

Chapter 2 analyses the intense struggle that took place as the dominant classes diversified into different fractions with increasingly different interests in the development of capitalism. I argue that the main antagonists during this period were not capital and labour, since wage labour was not highly developed, but the differentiated industrial, financial, commercial and agricultural interests. These struggles intensified as their goal became nothing less than to gain a monopoly of the control of state power. The Second World War marked a decisive point in these struggles, as the war greatly expanded the export market for agricultural products, and led directly to the forging of an alliance between agricultural and commercial capital. This alliance was strengthened by support from the state bureaucracy. Its power became evident when the Democrat Party, representing the interests of the agricultural-commercial capital alliance, humiliated the Kemalist Republican People's Party in the 1950 elections. The RPP had been the founding political organ of the Turkish Republic, and the agent of etatism. It represented the interests of industrial capital, with popular support from the peasants who responded to its nationalist, Kemalist ideology.

The concluding argument of the chapter is that the 1950 election marks the start of a new period in Turkish history in which shifting alliances were made between fractions of capital, but in which no one fraction managed to gain an effective monopoly of state power. Although the agricultural-commercial capital alliance during and after the Second World War, established its domination over state power, it could not prevent a fierce and determined challenge from industrial and financial capital. The resulting instability and struggles have led to forms of politics and state organisation which have prevented the smooth transition from a closed to an open economy. It is here that we find the roots of the economic and political crises and military interventions of 1960, 1971 and, most importantly, of 1980.

Chapter 3, The Politics of Industrialisation in a Closed Economy and the IMF Intervention of 1979, by Huseyin Ramazanoglu, deals with the development of Turkish capitalism in a closed economy and the political problems which are generated in such an economic system, especially when the processes of capital accumulation are blocked by the political and ideological structures. In this chapter, I also examine the crucial role played by the IMF in forcing Turkey to open its economy to world markets, thus irrevocably affecting the present as well as the future development of the Turkish social formation.

The Turkish experience of capitalist development has been a very complex process, which has proved highly resilient to a variety of threats to the system. The long Ottoman experience of building and running the Empire must have contributed to the development of political practices on the part of the dominant classes, which resulted in Turkey being relatively unknown in the rest of the world, and also becoming the object of prejudice. This could be due, at least in part, to the imperial mentality of the Ottomans, and the survival of this mentality in the Republic of Turkey, however inappropriate this might seem. Cultural isolation from the outside world led in this century to a desire to imitate the West without really understanding it.

Capitalism as a world system imposes uniformity, and this creates resistance in many developing countries. Turkey was relatively sheltered by its isolation from this tendency of world capitalism to penetrate and alter societies and their cultures, while its own capitalist system was being established. The contradictions inherent in the policy of isolationism were sharpened when it became necessary to move towards an outward-looking economy. Initial attempts to open the economy, and thus the country to the rest of the world aroused fierce opposition, on the grounds that Turkey did not need foreigners. This response is still characteristic of the post-1980 military government, and is closely related to the problems of establishing democracy in Turkey. The military leaders, especially General Evren, retain an intransigent nationalist attitude in the face of Western criticism of the regime's dubious human rights record, but this an attitude which enjoys considerable popular support. This nationalist isolation has perhaps concealed the fact that Turkey has always been a staunch ally of the West, implacably opposed to the emergence of any form of socialism, and an integral part of the world capitalist system.

The political situation which emerged over a period of three decades (1950-80) has been contorted and, at times very unstable. There have been three coups and various attempts at representative democracy, with coalitions, minority governments and protracted parliamentary crisis all playing their parts in shaping the transformation of the Turkish social formation. The Turkish case provides an excellent example of the kind of difficulties that newly industrialising countries face today where the contradictions of capitalist development are so obvious and yet their resolution so difficult. The accumulation of capital, as the primary aim of capitalist reproduction, has been so much at the mercy of political experimentation and the threat of international political opinion, that it is sometimes hard to believe that Turkey has so long maintained a closed economy and yet managed to be influenced so strongly by external pressures.

This chapter analyses the struggle for the monopolistic use of state power between the fractions of the Turkish bourgeoisie and show that it was the strong etatist nature of the state and state policies which prevented the successful outcome of this struggle in favour of the dominant fractions of the bourgeoisie, i.e. monopoly/industrial and financial capital, within the framework of the parliamentary system. The first two military coups (1960 and 1971) were attempts to restructure the Turkish state without undermining its etatist base. The third one of 1980, however, was completely different, in that it was aimed at specifically displacing that base and opening the economy to world markets, with new emphasis on export-promotion, and a substantially revised state structure geared for the maintenance and the reproduction of these new structures and relationships. Analysis of the 1980 coup, and the attempts to open the economy to world markets is undertaken in Chapter 8, but here the focus is on the examination of the politics of industrialisation in a closed economy, and the internal and external strains which have led to the downfall of Turkish democracy and its replacement with military rule.

The need for a change of direction in the development of Turkish capitalism caused a considerable stir among social scientists and other intellectuals in Turkey, with controversies raging over competing explanations of the economic crisis and over appropriate strategies for resolving the crisis. Kutlay Ebiri in Chapter 4, Turkish Apertura, provides a very useful analysis of the main issues in this debate.

Ebiri relates the debates over the opening of Turkey's economy to the discussion of similar issues in other countries at comparable levels of capitalist development. He identifies two main parties to the debate. There are those who accept the assumptions of the dependency school, and who thus oppose the opening of the economy to world markets on the grounds that further imperialist penetration would exacerbate the problems of underdevelopment and dependence in Turkey. Others reject these assumptions and argue that since capitalism is so deeply rooted in Turkey, and the social formation itself is structured by capitalism, its development cannot be curtailed and the economy should be opened. The debate is not simply over whether Turkey should shift from a closed to an open economy, but also over whether Turkey can withdraw from the world capitalist system and follow a socialist path of development. Turkish adherents of the dependency school on the whole take the view that Turkish nationalism should be used against the dependent development created by imperialism, allowing Turkey to break away from the capitalist system, and its international conspiracy to subordinate Turkey, while those who reject dependency views feel that nationalism does not necessarily lead to socialism, and that there is no international conspiracy,since Turkey is an integral part of the world capitalist system.

Ebiri points out that Turkish dependency theorists use the assumption that Turkey has been and is being underdeveloped, as part of an international capitalist conspiracy to argue against the opening of the economy. Arguments for the opening of the economy are, therefore, based on quite different reasoning, and in particular on the failure of import-substitution as a development strategy. The underdevelopment/conspiracy argument suggests that had import-substitution been more carefully planned, and the policies more moderate, Turkey could have ridden out the world recession. The alternative view, which is the one favoured by Ebiri, attributes the cause of the crisis to the use of import-substitution as a long term strategy. This view rejects the assumption that there was a right path which import-substitution could have taken, and argues that given the specific conditions of the Turkish economy and Turkey's social and political relations, there was no alternative to developing an initial period of import-substitution, during which a domestic consumer goods industry could be established. But when the next step, to establish a capital goods industry, had to be taken, the structure of Turkish industry was weakened rather than strengthened.

Ebiri also draws attention to two special advantages enjoyed by industry during the import-substitution period. The first one was the lack of effective centralised control of the economy, which created imbalances in the development programme. The second was the massive flow of remittances from Turkish workers in Europe, which cushioned the downward trend of the economy, and postponed the crisis until after 1974, when the impact of world recession hit the inward-looking Turkish economy particularly badly.

Ebiri argues forcefully that Turkey cannot become more competitive in world markets without a complete shift in economic strategy. This shift would entail not only the opening of the economy, but also an increase of technological input into production, particularly in those sectors of the economy where export promotion is most feasible. He concludes that Turkey does not have a choice between a "smoothly functioning" import-substitution strategy and an export-promotion strategy. Import-substitution is no longer an effective basis for development, and it is essential to open the economy if Turkey is to continue to develop. Ebiri is, however, aware of the ideological and political difficulties that this shift in strategy will bring.

Fikret Senses takes up some points from this debate in Chapter 5, Short-Term Stabilisation Policies in a Developing Economy: The Turkish Experience in 1980 in Long-Term Perspective. He elaborates a number of issues arising from the long-term effects of the stabilisation policies which were adopted on 15th January 1980. He agrees with the other contributors to this volume in seeing the onset of world recession in the 1970s as a critical point in Turkey's development. He argues that Turkey's import-substitution strategy resembled the policies of a number of other developing countries during the 1960s, but this policy was geared to the production of consumer goods, and could not easily be extended into intermediate and capital goods. The crisis intensified as import-substitution failed to earn sufficient foreign exchange. The end of import-substitution does not mean the end of imports, and the lack of foreign exchange earnings to pay for the import of essential intermediate and capital goods imposed a critical restriction on the economy.

Senses goes on to examine briefly the long-term effects of import-substitution, and the various attempts which were made to stabilise the economy between the onset of the recession in 1973 and the adoption of the stabilisation programme in 1980. He also looks at the objectives of the 1980 policy programme, and examines its impact on selected economic indicators. He concludes that in spite of its initial success in bringing down the inflation rate, the simple reliance on monetary control and market mechanisms advocated in the 1980 stabilisation programme could not be expected to rescue Turkey from its current difficulties. In addition, the measures needed to achieve self-sustaining growth and to eliminate obstacles to growth, would undoubtedly generate determined opposition from vested agricultural and commercial interests.

It is apparent from these analyses of Turkey's economic crisis, that the future development of Turkish capitalism depends on the ability of the government of the day to restructure the framework of economic and political relations. This restructuring, however, is extremely problematic. Opening the Turkish economy must involve reallocating resources from agriculture and commerce into industry, which entails a drastic shift in the balance of power within the dominant classes. The military regime recognised that it had formidable problems in overseeing the upheavals that would necessarily ensue, but if it could not ensure the development of manufacturing industry that could compete in world markets, the opening of the economy could not succeed.

The essays by Ebiri and Senses comprise the analysis of the crisis in the Turkish economy, and they raise a number of important questions about the need to open the economy and about the similarities and differences between the situation in Turkey and those elsewhere. A major omission from the arguments summarised so far is the impact on the development of Turkish capitalism of the accelerated proletarianisation of the Turkish peasantry since the 1960s. The relation of this process to labour migration as a general issue and to the crisis of Turkish capitalism is dealt with by Caroline Ramazanoglu in Chapter 6, Labour Migration in the Development of Turkish Capitalism.

This chapter gives a brief overview of the debates on the place of labour migration in the development of capitalism. This is a general theoretical critique of the dependency theory of labour migration as a typical process by which central capitalism more or less mechanically produces reservoirs of cheap labour in the periphery, and of the articulation of modes of production view that labour migration can be analysed as a general phenomenon in the development of world capitalism.

The argument is made that Turkish labour migration can only be understood if the transformation of pre-capitalist relations of production in Turkey, and the contradictions which have developed in this transformation, are taken into account. The development of labour migration within Turkey and from Turkey cannot usefully be seen as a typical (or atypical) Third World process, as it has specific historical causes and consequences. It is undoubtedly true that there are similarities between the kinds of internal and international labour migration which have developed in Turkey, and the kinds of labour migration which are found elsewhere, but these similarities (and also the differences which can be found) need to be explained, and not taken for granted in the construction of a theory.

Turkish labour migration differs from migration elsewhere in that it is closely related to the development of class struggles and the state in Turkey, and thus to the crises of Turkish capitalism, but since these upheavals occur in the context of the world capitalist system, Turkish labour migration is also part of the general development of world capitalism and needs to be understood at a general level. In this chapter, therefore, the case of Turkish labour migration (drawing in part on her fieldwork in Istanbul) is used to elaborate and illustrate the argument that explanations of any form of labour migration must incorporate analysis both of the specificity of the social formation in which relations of production are being transformed, and of the totality (the world capitalist system) of which the social formation in question is a constituent part.

This theoretical position is supported by looking briefly at the development of labour migration in Turkey in relation to the transformation of relations of production, and also at the intervention of the state in the processes of class formation which have given rise to mass movements of labour.

As an example, the combination of circumstances which resulted in the end of a successful period of import-substitution, coinciding with the deepening of the recession in Europe, in the 1970s is examined. This complex coincidence had a direct impact on the development of labour migration, and led to a dramatic intensification of class struggles in Turkey, in which the development of labour migration was a conditioning factor.

The chapter concludes with the argument that labour migration cannot be treated as a common stage in the development of capitalism which is independent of the development of specific capitalist states, classes and their various struggles. The transformation of small direct producers into workers, or seekers of work, has become an extremely rapid and inflexible process, but although it occurs in many parts of the world, it is a process which is incomplete, and one which is not easily predictable. This fundamental transformation of production relations is not a simple process of transition to a known end; it contains major contradictions, whose resolution can take different forms at different historical moments.

Turkey's policy of exporting workers was not a simple response to the needs of central capitalism for cheap labour, but a short term and desperate measure which indicated the state's loss of control over the transformation of agriculture, the failure of the bourgeoisie to develop intermediate and heavy industry, and the resulting pauperisation of a large proportion of the population.

The theme of struggle within Turkey is taken up by Feroz Ahmad in Chapter 7, Military Intervention and the Crisis in Turkey. Ahmad examines the two decades (1960-80) in a careful study of the economic and political relationships between conflicting interests. In this period Turkey had one of the highest rates of growth in the world, due primarily to the vast size of the domestic market, and its ability to absorb and consume goods produced by the rapidly growing industrial sector. This rapid industrialisation had also given rise to the emergence and development of a large and well organised industrial labour force. Ahmad charts the growing influence of the working class on the development of Turkish capitalism.

This essay starts by examining the military intervention of September 1980, and continues by analysing events immediately prior to the coup. This directs his attention to the relationship between Turkish and international capital, and their respective gains from the coup. He argues that the coup was inevitable as there was no other way in which political obstacles to the further accumulation of capital could be removed. This argument provides the framework for the rest of the essay, in which the focus is on the complex and highly antagonistic nature of Turkish party politics, and on the problematic relationships between political struggles and economic development.

Ahmad's analysis takes him back to an examination of the genesis of the crisis. He looks briefly at the Democrat Party rule, from 1950 to its overthrow by the first military coup in 1956, and then at the industrialisation programme which was implemented after the 1960 coup when the Republican People's Party returned to power. He continues with an analysis of the politics of industrialisation, where he explains very carefully that the creation of Turkish democracy was incompatible both with the demands of industrialisation, and with the class interests of what he terms big business (which might alternatively be conceptualised as financial and industrial capital). He also examines conflicts within big business over whether the tensions in the system could be relieved by the development of social democracy. He concludes that although Ecevit, as leader of the RPP governments in 1973 and 1974, attempted democratic rule, these attempts were ineffectual.

Ahmad disentangles the causes of Ecevit's difficulties, and brings out in particular, not only the political obstacles which were put in Ecevit's way, but also the unrealistic strategies by which Ecevit tried to locate Turkey as part of the non-aligned Third World, and to resist intervention by the IMF. He sees Ecevit's main political opponent, Demirel, as much more realistic in recognising that the needs of big business had to be met. Demirel was willing to accept the complete IMF stabilisation package, but he was prepared to condone fascist practices and the destruction of democracy as part of the price to be paid. The measures taken by the military since 1980, and the implications of these measures for the future development of Turkish capitalism within the world capitalist system are taken up in the following chapter.

Chapter 8, The State, the Military and the Development of Capitalism in an Open Economy, by Huseyin Ramazanoglu, is an examination of the specific measures taken by the Turkish military regime between 1980 and 1983, together with a more general analysis of the relations between the military regime and the state. The military regime embarked on the wholesale reorganisation of political, economic and ideological structures after 1980, and irreversible policies were rapidly implemented, with many further changes planned. Although it is far from clear as to whether this regime and its class allies can actually achieve all their goals, we can realistically expect capitalism to develop in Turkey in a way that is substantially different from the form it has previously taken.

Although the military government appeared to be firmly in control, there were multiple undercurrents which could disrupt the smooth transformation of Turkish capitalism form an inward-looking to an outward-looking economy. Not least of these were the divisions between the military leaders themselves, as the balance of decision-making gradually tilted towards the extreme rightwing elements within the National Security Council where all major decision were taken.

There are obviously very close connections between what financial and monopoly/industrial interests see as the current economic pre-requisites of Turkish capitalism, and the new political structures which were slowly being planned and established in correspondence with these pre-requisites. This essay is intended to show this close and intricate relationship, in the specific case of Turkey, and also to consider its more general significance.

This chapter concentrates on a political evaluation of the economic measures taken by the government, and in particular on the effects of these new and significant developments on class struggles in Turkey. I analyse the available channels of political representation, the changing nature of political power, and the framework and scope for political activity, on the assumption that an analysis of class struggle in a social formation can most fruitfully be studied by examining the organisational structures of the state. It is in these organisational structures that any level of class struggle is crystallised, and this is where direct access to the distribution of political power can be gained or denied. This brief analysis is intended to clarify short but critical period in the reorganisation of the Turkish state, and to show that these events are relevant to the analysis of class struggles elsewhere.

Briefly considered is the nature of the military's efforts in establishing a "democratic" political framework in Turkey, and the question of whether or not these political measures could lay the foundations for a democratic framework of political representation. The results of the 6th November 1983 elections are briefly considered in order to evaluate the chances of the Ozal Government being able to meet their policy objectives. There seems to be a great deal of truth in Lenin's celebrated passage that,

> "A democratic republic is the best possible political shell for capitalism, and therefore, once capital has gained possession of this very best shell...it establishes its power so securely, so firmly, that <u>no</u> change of persons, institutions, or parties in the bourgeois-democratic republic can shake it."
> (Lenin, 1970, p.296)

This is indeed the scenario that has always been played in Turkey. The viability of the Turkish capitalist state and the maintenance of the processes of capital accumulation necessitate continuous attempts at experimenting with various types of democracy.

In these general and specific analyses of the problems of shifting from a closed to an open economy, we have to bear in mind that the selection of any economic strategy is a political matter. Any economic policy decision is the end result of a political struggle, and the process of choosing between alternative economic strategies is a process imbued with class contradictions. The cutting of this Gordian knot in 1980 by military intervention was resorted to only after attempts to bring about the necessary reorganisation through parliamentary means had failed. In the absence of any well organised leftwing movement, the coup of 1980 can be seen as the culmination of a process in which the dominant classes unsuccessfully attempted to resolve their struggle for power and domination. The situation in which financial and monopoly/industrial capital were the dominant economic power, but could not monopolise state power, was too unstable to be allowed to continue.

The Turkish left has been undoubtedly most adversely affected by the struggle for monopolistic use of state power between the fractions of Turkish capital. It has suffered heavily as a result of adopting an analysis of capitalist development which prevented them from having a clear understanding of class struggles and the organisation of the state within Turkey. As a result of a preoccupation with imperialism and underdevelopment, the left has been caught up in the struggle between the fractions of capital without achieving a clear conception of what it was up against. The real tragedy of the left is that while it has little hope of re-grouping effectively in the near future, it could probably have become an organised force to be reckoned with, under an efficient and democratic capitalist system.

One reason why the left has developed such a limited understanding of class struggles in Turkey is the quality of existing analyses of Turkish society and of capitalist development in general. While there are many eminent Turkish social scientists and many foreign scholars who have published on Turkey, these studies are primarily in the tradition of modernisation approach, or more recently, are of the dependency school. Neither of these schools contribute to the fulfillment of the need for conjunctural analysis of the transformation of the Turkish social formation, which is sadly lacking in Turkey. It is only recently that analyses aimed at the examination of the specificity of the Turkish social formation are beginning to develop, and at this stage not all the generalisations that we need to make can be substantiated with existing empirical work.

While Turkey has been locked in its own internal struggles, the need to open the country to the outside world has become a pressing matter not only within Turkey, but also from the perspective of international capital. Turkey's strategic position in NATO and in relation to the Middle East. Its growing domestic market makes internal upheavals of direct interest to the centres of international capital in the world system. Turkey, like any part of the system, can no longer be allowed to go in its own way to perdition. Turkey's struggle to survive in a rapidly changing world must be seen as a matter of general concern.

REFERENCE

Lenin, V.I. (1970); "State and Revolution" in Selected Works, Vol.2, Progress Publishers, Moscow

1 Current problems in conceptualising the uneven development of capitalism [1]

HUSEYIN RAMAZANOGLU AND CAROLINE RAMAZANOGLU

In the post-war period, social scientists of differing theoretical and political persuasions, have looked for ways of conceptualising the expansion of capitalism into the so-called Third World. In a period in which rapid industrialisation is taking place in most parts of the world, in which the differentiation between the immense wealth of those who have and the mass starvation of those who have not, is accepted as unremarkable, theories of development have had surprisingly little to offer by way of explanation and alternatives, in spite of the vast quantity of literature which has been produced. Social theories of development have taken three main directions and in this chapter we briefly review the main assumptions of the modernisation school, the dependency school and the debate on the articulation of modes of production, before going on to consider the nature of the world capitalist system, and to suggest more useful ways for accounting for the complexities of the development of world capitalism.

POST-WAR THEORIES OF DEVELOPMENT

1. Modernisation theories

In the immediate post-war period, the obvious differences in the levels of development of different parts of the world were conceived in terms of degrees of modernisation. A variety of social, political and economic theories classified levels of development according to the ability of local populations to overcome internal obstacles to modernisation and to reach more advanced points of economic take-off and political democracy. The United States of America and Western European societies represented levels of specialisation, modernity and economic development which the rest of the world had unfortunately failed to achieve. Although the rest of the world was seen as in a process of transition up an evolutionary scale, lack of development was attributed to isolation from contact with the advanced nations or, where contact was recognised, to the resistance of traditional value systems and social structures to modernising ideas and agencies

2. Dependency theories

From the late 1960s, these theories were seriously undermined by the rise of theories of dependency and underdevelopment which challenged this ahistorical and ethnocentric analysis. During the 1970s, these dependency theories (theories of underdevelopment, unequal exchange and world system) became the dominant means of accounting for the expansion of capitalism into the Third World and the uneven levels of development which could be observed.

Whereas modernisation theorists conceived lack of development in terms of failure to develop internal specialisation, appropriate values and other pre-requisites for growth, dependency theory, conceived underdevelopment as a process in which economic growth had become stunted and restricted by the rise of the capitalist mode of production. Dependency theory is based on a fundamental distinction between the original transition to capitalism which was successfully achieved by much of Europe, the United States of America, and some other economies, and the incomplete transition to capitalism which has occurred in the rest of the world. The typical transition to capitalism achieved by the centre, core or metropoles of capitalism, is contrasted with the typical transition to distorted peripheral capitalism in the Third World, periphery or satellites. These typically different paths of development are seen as necessarily dependent on each other in that the accumulation of capital in the centres of developed capitalism can be shown to have been dependent on the extraction of various forms of value from the peripheral areas. Underdevelopment is not, therefore, a matter of slow progress up the evolutionary scale in which some try harder than others, but an active process of exploitation in which the whole of the Third World has become locked into an international system of unequal exchange.

The emphasis in dependency theory has never been on the ways in which underdeveloped economies or societies have differed from each other, although there is some discussion and debate in the literature. Theories of dependency are primarily ways of accounting for a common process of underdevelopment in which exploitation within the world system of capitalism leads to typical internal patterns of development in the Third World. Dependency theorists argue that countries with very different pre-colonial or pre-imperial histories begin to look very similar in terms of their disadvantaged position in the world market, their active exploitation by the centres of advanced capitalism, and their typically underdeveloped internal economic and social structures. The rigidity of dependency theory is shown in the belief that development is never possible once the typical features of peripheral capitalism have developed, since further development can only be the development of underdevelopment. Since there is clearly development in the so-called periphery, the more industrialised peripheral social formations are characterised as "semi-periphery", and as mediating between the "core" and the "periphery". This conception does not really solve the problem of explaining high levels of development achieved in the Newly Industrialising Countries of the world, but instead emphasises further the difficulties which are inherent in this approach.

While dependency theorists effectively exposed the weaknesses and implicit politics of modernisation theory, and dramatically demonstrated through colonial and imperial history the incorporation of the entire world into the world capitalist system, their accounts of the expansion of capitalism are seriously flawed by a number of unwarranted assumptions.

(i) While criticising the universalism of modernisation theory, dependency theorists have generated a comparable abstract, universal and thus, ultimately, ahistorical theory. Although dependency theory has been valuable in emphasising that all "traditional" societies have histories, its theorists have failed to look at the different ways in which the capitalist mode of production has transformed pre-capitalist social formations. By having to classify all cases as either typical or atypical of peripheral capitalism, they have had to overlook historically critical differences in the way in which pre-capitalist social formations have been incorporated into the world capitalist system. They have not, however, been able to justify this historical determination, and this implicit assumption in their work seems best explained as deriving uncritically from Lenin's pamphlet on imperialism, primarily through the influence of Baran and Sweezy on Frank (Lenin, 1970; Baran, 1957; Baran and Sweezy, 1968; Frank, 1969, 1971, 1972, 1975, 1978a, 1978b, 1980, 1981). Lenin seriously underestimated the diversity of possible paths of development which could result from differences in pre-capitalist productive systems and the variable ways in which they could be incorporated into the world capitalist system.

(ii) The universalism of the theory is justified by the additional assumption that capitalism on a world scale is a system of exchange rather than interrelated systems of production, circulation, distribution and exchange. In the dependency view, it is the inequality and exploitativeness of exchange relations dominated by the centres of capitalism which prevents the Third World from developing. This is the second point which seems to derive from uncritical acceptance of Leninist orthodoxy and which has forced dependency analysis into a position which must deny the possibility of development in the Third World. Dependency theorists have indeed documented patterns of exchange and exploitation which did not characterise the original transition to capitalism, but they have failed to justify the generalisation of these histories into a theory of historical determination.

(iii) If these two assumptions are left unquestioned, it follows quite logically that the contradictions of the development of capitalism lie within the international system of exchange relations rather than in the production systems of different social formations. Dependency theorists differ somewhat in the emphasis they put on the analysis of class and state, but they agree that class antagonisms and thus class struggles are located in the exploitation of the periphery by the centre. Class analysis, therefore, looks at the opposed interests of the international bourgeoisie with its Third World allies, and the (admittedly differentiated) workers and peasants who make up the bulk of the world's population. Wallerstein puts forward one of the crudest versions of this view when he argues that while in core areas national, class conscious struggles for control of the state do occur, in peripheral areas class struggle can only take the form of national liberation struggles in which the local bourgeoisie will always be allied with international capital (Wallerstein, 1974, 1979, 1980, 1982). This view, incidentally, makes it quite impossible to understand recent Turkish history, the tremendous struggle within the Turkish bourgeoisie to gain control of the state, and the 1980 coup; see Chapters 2, 3, 7, 8 of this volume. It follows from this view of class that the development of states in the Third World are given little attention since they must take standard peripheral forms.

(iv) Since dependency theory is located within a broadly Marxist social theory (although not all dependency theorists see themselves as Marxists) prescriptions for the achievement of socialism characterise the literature. Given the above framework of assumptions, dependency theorists see the overthrow of international capital by Third World workers and peasants as unrealistic and, therefore, recommend national liberation struggles which would enable peripheral social formations to withdraw from the exploitative exchange relations of the world system of capitalism, and to pursue some form of independent development.

The view of capitalism as a central parasite preying on its own periphery is only logical if the other assumptions of dependency theory are accepted. The idea that all relationships within the world system of capitalism are similarly exploitative and parasitic seems unnecessarily unrealistic. The so-called Third World is highly differentiated and we need to explain the differences in development between states such as Chad and India for example, but also the internal differentiation and internal struggles within states. Uneven development can be observed in every part of the world system, and it no longer seems useful to try and force varied histories into a straitjacket of dependent development. There is undoubtedly a great deal of exploitation which can be documented. but we would argue that there is also development and, in order to understand the reproduction of the world capitalist system, we have to be able to conceptualise development at the level at which it occurs, that is in individual social formations. To argue that capitalism must generate underdevelopment and must, therefore, be resisted in favour of independence is to misunderstand recent history and to misrepresent the operation of world capitalism.

3. The debate on the articulation of modes of production

The eclipse of modernisation theories by the dependency school redirected the conceptualisation of the expansion of capitalism, but left central problems of interpretation unresolved. More recently as dependency theory itself has come under attack, subsequent discussions have focussed on the articulation of modes of production; a debate over the best way of conceptualising the logic and consequences of the expanded reproduction of the capitalist mode of production as the entire world becomes incorporated into the world capitalist system. [2]

The chief participants in this debate differ over the way the key issues are viewed, but there seems to be general agreement that, as in dependency theory, the process under discussion is a generalised transition to capitalism which, in different ways, the Third World or "peripheral capitalist" social formations have failed to compete. [3] In the articulation debate, this incomplete transition supposedly takes the form of the problematic co-existence, or articulation, of two or more of modes of production in a given social formation, one of which is capitalist, dominant and expanding, while the other (or others) is pre-capitalist, subordinate and simple in reproduction, and thus ultimately doomed to extinction. These ideas seem deceptively simple in spite of the verbal complexity with which they are usually presented, but the debate takes place on a level of such abstraction that an immense degree of disagreement is possible over every point, with no clear prospect that such disagreement can be resolved by reasoning or general principles.

The debate on the articulation of modes of production has served to clarify which issues in the conceptualisation of the development of capitalism remain problematic, but its basic premises are misconceived and unnecessarily limiting. The debate has arisen within a particular framework of assumptions which has resulted in unwarranted emphasis being put on an assumed general transition to capitalism, while inadequate attention has been paid to the relation of general historical processes to concrete, historical situations, and the real struggles in which people are actually engaged. In this chapter, we want to look again at the general problems of conceptualising the uneven development of capitalism on a world scale, but with the intention of understanding general historical processes of capitalist development as they relate to specific, historical situations.

We want to emphasise here that we do not mean that all social analysis should be restricted to the concrete and unique, but that while the logic and consequences of capitalist development may be the same at a general level of abstraction, through history different forms have emerged which are specific to a given social formation at any particular period. A theoretical understanding of the uneven development of capitalism must recognise the relationships between general processes and historical forms. This point is not a new one, but it is our contention that the implications of taking this position have become lost in the confusions of the dependency and articulation literature and, in particular, that the relationship between general processes and historical forms remains problematic. While general processes of capitalist development occur in every capitalist social formation, each historically determined social formation develops unique forms of class struggle in a unique capitalist state, and a unique set of relationships with the rest of the world capitalist system, which cannot be understood either as general/typical Third World forms, or as isolated and atypical aberrations. They can only be understood as unique parts of the whole capitalist system. Before discussing the implications of taking this view, however, we need to clarify our position in relation to the articulation debate, and to propose an approach to the uneven development of capitalism which can take the operation of the whole and its parts into account. We look first at the general theoretical problems raised by the articulation debate.

THEORETICAL PROBLEMS ARISING IN THE DEBATE ON THE ARTICULATION OF MODES OF PRODUCTION

1. Basic premises

The articulation debate cannot easily be summarised, since the central concepts and issues are by definition contested, but the following points can be taken as its basic premises.

(i) The debate can be seen as a reaction to the theoretical and methodological assumptions of the dependency school and, in particular, as a turning away from the conceptualisation of history as a universal process, with development necessarily creating underdevelopment and *vice versa*. The existence of gross inequalities around the world is acknowledged, but the explanation of these differences given by dependency theorists is challenged. Participants in the debate generally reject the mechanistic conception of central capitalism parasiticly preying on its own peripheral social formations. They attempt to return to a more classically Marxist theory and method in approaching the problems of interpretation, but there is no unanimity in the manner of their rejection, and no agreement on what exactly constitutes Marxist theory and method.

(ii) There is, however, agreement that the return to Marx means distinguishing between the general abstract processes which characterise a mode of production in general, and the variable historical realities which can actually be found in different capitalist social formations. Those engaged in the debate are united in their criticism of the dependency school for failure to distinguish between mode of production and social formation in this way. They argue that capitalism cannot be generalised into a single universal mode of production, without regard to the variable ways in which pre-capitalist modes of production can persist and co-exist with the capitalist mode. The participants in the debate, however, disagree over how this essential distinction should be made, and on the implications of making it. In our opinion, the conceptualisation of the history of capitalism as a process of transition, and the use of argument at a generalised level of abstraction has led to a general failure to follow through the logical consequences of distinguishing between mode of production and social formation.

(iii) The value of distinguishing between mode of production and social formation is, nevertheless, a central assumption in the debate and it is agreed that modes of production can only be said to be realised in actual, historical, social formations. Articulation then becomes a key concept because it is used to identify different forms of co-existence between pre-capitalist and capitalist relations of production which can develop in social formations during the assumed transition to capitalism. The advantage of the concept of articulation is that it enables us to look beneath the blanket notion of underdevelopment, and to ask what happens to pre-capitalist relations of production when the capitalist mode of production develops in a social formation; what forms of co-existence then develop between the modes of production; how these might be conceived; the implications of these forms, and how and why they vary. The weakness of the concept is that it confuses analysis of particular situations with assumed general processes. In spite of the advantages of the concept, the whole debate has become hamstrung by the assumption of a transition to capitalism which sets pre-determined limits on the possible forms of co-existence which can develop.

(iv) This limitation has led to a situation in which participants in the debate castigate dependency theorists for their failure to take account of the historically specific ways in which the pre-capitalist modes of production become articulated with the capitalist mode, but they themselves then tie specific conjunctures to an assumed general process of transition. This means that while key questions can be asked about the development of capitalism, there is no agreement on how they might be answered. This lack of agreement leaves unresolved crucial questions of how pre-capitalist modes of production can be identified and classified, how given modes become able or unable to maintain their conditions of existence, how they can become subordinated, and why different forms are variously destroyed, transformed or able to persist.

(v) It is generally agreed that by its nature the capitalist mode of production cannot exist harmoniously with pre-capitalist modes, because there are contradictions between the conditions for the expanded reproduction of the capitalist mode of production and the conditions for the simple reproduction of the pre-capitalist modes, but the outcomes of these contradictions remain disputed. The debate, therefore, continues inconclusively over what forms of development, as opposed to underdevelopment are possible at the level of social formation.

(vi) Recent contributions to the debate by Wolpe and Alavi, and an earlier intervention by Leys and Foster-Carter, indicate the problems in the level of argument adopted, since the debate is over general propositions and the search for general solutions to the problem of what happens to pre-capitalist relations of production in particular social formations as capitalist relations of production rise to dominance (Wolpe, 1980; Alavi, 1982; Leys, 1977; Foster-Carter, 1978). The only apparent justification for taking this general line of reasoning seems to come from assuming a general transition to capitalism, and it is to this assumption that we now turn.

2. The transition to capitalism

The complexity of the subject matter and the ambiguity of key concepts, notably mode of production, social formation, articulation and class, make the clear expression of the passage from pre-capitalism to capitalism a formidable task. The theoretical study of the development of capitalism has conventionally been based on the study of the original transition from feudalism to capitalism in Europe. Authoritative works on the subject, however, provide various interpretations of this transition, with necessarily different theoretical and political implications. [4] Generally, research on the problem of the demise of the feudal mode of production and the emergence of the capitalist mode has taken the transition to be a linear progression with the former giving rise to the latter. The literature on dependency, and on the articulation of modes of production is imbued with this concept of transition, which limits possible understanding of the diverse ways in which capitalism can expand and develop. There does not seem to be any obvious justification for generalising European experience in one epoch into a general theory of capitalist development. Although at one level any development of capitalism is part of the reproduction of capitalism as a world system, capitalism does not have to develop out of feudalism, and so the conditions of reproduction today can logically and in practice take different forms; "European" or "Western" capitalism itself cannot be taken to have been a uniform historical experience.

This unjustified assumption then, seems to be a fundamental problem with both dependency theory and the articulation of modes of production debate, since key assumptions are derived from analysis of conjunctures in European history when the capitalist mode of production had already asserted its dominance. The concept of a general transition to capitalism them becomes a set of assumptions which can only be understood after the linear process has been completed. We propose that this concept of transition has little utility for interpreting the more recent history of capitalism, and that it would be better to adopt as a conceptual tool the notion of transformation. [5] In suggesting the use of transformation rather than transition, we are suggesting a methodological shift in the study of the extremely complex situations in which the capitalist mode of production expands at the expense of the non-capitalist mode in particular social formations, and also the rejection of a set of assumptions which actively hinder this study.

By transformation, we mean the process whereby different social relations of production in a given social formation are transformed so that in due course one set of social relations of production establishes dominance over the rest, but without this dominance necessarily being achieved at the expense of total dissolution of previous forms of social relations of production. This means that we do not have a priori grounds for expecting uniformity at the level of social formation; logically we should expect some degree of variation. It is possible that transformation is not ideal term for our purpose, but we use it here to draw attention to the simultaneous processes which we are trying to identify theoretically and empirically, and to the methodological problems that ensue if we abandon the expectation of common forms of "underdevelopment" or "articulation" in a Third World. While transition denotes a linear development of social formations, transformation denotes complex processes taking place simultaneously, whose consequences are not predetermined. The concept of transition is necessarily limiting, whereas the concept of transformation makes it possible to begin to conceptualise not only the development of social relations of production at a general level, but also at the level of discernible class practices, and to identify the distribution of political power at various phases of capital accumulation. The overall advantage of the concept of transformation is that it enables us to specify the levels at which crucial variations in the development of capitalism take place.

The real complexity of the variations which can be found in different social formations cannot usefully be generalised without much more careful analysis of why variations occur. In particular, they cannot usefully be squeezed into the framework of a linear view of history. Capitalism today has developed to a level where it governs the international division of labour, but this appearance of a world system of capitalism should not be used to obscure the different forms which capitalism takes. One way in which these forms continue to be obscured is by the retention of terms such as Third World and "peripheral capitalism", even when the dependency theory which gave rise to then is rejected.

3. The problem of Third World as an analytical category

The dependency theorists are criticised for generalising capitalism into a world mode of production, but the assumptions of the articulation debate lead to a contradictory position where dependent development terms, and the assumption of the uniformity of capitalist development perversely creeps back in. The use of terms such as Third World and "peripheral capitalism" effectively obscure crucial variations in development chiefly because they do not permit analysis of the development of specific class struggles at particular periods, or of different roles played by different states at different historical conjunctures.

The problems of combining a rejection of dependency theories with the continued use of general concepts of Third World and "underdevelopment" are shown in Taylor's attempt to provide a textbook based on articulation premises. Taylor clearly argues for the existence of common Third World structures of underdevelopment, with their economic base in an articulation process. This presents him with the problem of specifying what the process is in particular cases, and the problem of identifying variations in the development of classes and states. Taylor gives an example which illustrates the problem. A study of Indonesia from 1880-1930 showed that labour struggles were divided between the economic struggles of the small permanent labour force in the extractive sector, and those of migrant labourers.

> "Whilst the struggles of seasonal and migrant labour were expressed in varying combinations of religious ideologies (syncretic forms of animism and Islam) produced in village units of the Asiatic mode, and realised in sporadic acts of militancy, the trained labour force in this sector tended to adopt forms of economic struggle more characteristic of those of prevalent in units of production in developed capitalist formations".
> (Taylor, 1979, p.238)

Taylor is only able to see this situation as an atypical example of differentiation of the proletariat during the colonial phase of imperialism. He is not able to analyse the specificity of the conditions which produced this situation, nor to develop the significance of of its uniqueness for the development of class struggles in Indonesia. Instead of being able to consider how far, and for what reasons there are comparable situations elsewhere, Taylor is forced into uneasy generalisations, with inexplicable exceptions to the rules imposed by transition. In order to avoid the pitfalls of this approach, we suggest that it is necessary to look much more carefully at the implications of making a distinction between mode of production and social formation.

4. Mode of production and social formation

There is a wide ranging literature on the nature of mode of production and social formation, and there have been some valuable attempts to clarify the concepts, (Althusser and Balibar, 1970; Godelier, 1972; Cook, 1977; Foster-Carter, 1978; Rey, 1979; Taylor, 1979; Wolpe, 1980), but while distinguishing between these concepts should enable us to embark on the study of uneven development, we do not think that we can advance our understanding much further unless we can apply the concepts to the analysis of actual situations. The transformation of pre-capitalist social formations can be conceived as a process in which the capitalist mode of production emerges in social formations previously dominated by other modes of production. The capitalist mode of production will then emerge as the dominant mode, and ultimately, will determine the nature of the social formation. The problem here is not to be able to state this as a general process, but to be able to identify the relevant factors which determine how this happens in particular instances, and what consequences then follow.

We take mode of production to be an abstraction indicating particular combinations of forces of production and relations of production. Relations of production denote the organisation and control of the forces of production. Modes of production can be distinguished from each other by their specific form of production of economic surplus, and also by their specific form of organisation of the means of production. Since mode of production is an abstract concept, it denotes relationships which can only actually exist in different, that is, historically specific, social formations. The capitalist mode of production then, designates a set of coherent structures within a social formation, and in particular, it designates a set of social relations of production and identifies classes that are embedded in these relations. The conditions of existence for the reproduction of the capitalist mode of production can only take place when these conditions appear in social formations. The specific character of each social formation, therefore, determines whether or not a given mode of production can be reproduced. At a general level, these processes will conform to the logic of the abstract framework, e.g. the development of capitalism, but there can be significant differences in the forms of capitalism that can actually develop in the variable circumstances of each social formation.

Mode of production is useful as an abstract concept in that it enables us to distinguish between different systems of production and social organisation, but these systems are created by people in the course of their daily lives within real, historical totalities. The term social formation refers to the configurations of specific class processes and other social processes - economic, political and cultural - which interact to determine, and be determined by each specific configuration. Each social formation is a totality which is always in the process of change. The unique character of each social formation can give rise to variations in the ownership of the means of production, appropriation or retention of the economic surplus, the division of labour, and the development of the labour process. The complex and interacting contradictions between the interests of the classes in a social formation provide the impetus for change, and change comes about with the development of various forms of class struggle. In different ways, the main form of struggle will be the efforts to suppress existing class practices in favour of the new class practices which can rise to dominance with the development of a new mode of production.

The participants in the articulation debate give great importance to this distinction between the abstract mode of production and different historical social formations, but much of their energy has gone into arguing, at an extreme level of abstraction, about the exact meaning of the concepts, [6] rather than using the distinction in order to learn what is actually involved in the complex transformation of pre-capitalism in actual social formations, although such attempts do exist (Bradby, 1975). It seems clear to us that these general conceptual disagreements cannot be resolved at the level of general abstract argument. The questions of which modes of production can co-exist with the capitalist mode, how their conditions of existence are transformed and how far their processes of reproduction are economic, political or ideological, are questions which can only be answered in their most concrete form. This means breaking away from the stifling debate on the concept of mode of production and instead, as has frequently been argued, locating the debate firmly at the level of analysis of social formations within the world capitalist system. [7] The implication of this decision is that further study will necessarily involve the use of empirical material from different social formations, and we can thankfully descend from the level of abstracted abstraction.

Since there are irreconcilable contradictions in the conditions of reproduction between the rising capitalist mode of production, and the various pre-capitalist modes in emergent capitalist social formations, the problems of transformation in these social formations can be conceptualised as the problem of securing the conditions of existence for the reproduction of the capitalist mode of production, in each social formation. This means that the reproduction of the conditions of existence of the various pre-capitalist modes of production necessarily becomes increasingly difficult, but not necessarily in pre-determined ways. We shall outline a general approach to these problems in the next section.

TRANSFORMATION: THE UNEVEN DEVELOPMENT OF CAPITALISM

Adopting the concept of transformation gives rise to formidable theoretical and methodological problems. In particular, there are very considerable problems, for which we may have no appropriate concepts, in actually studying how a given set of pre-capitalist social relations of production, with their structures of domination and subordination, become transformed into a different set of relationships with changed structures of domination and subordination. As Marx said, "the complete body is easier to study than its cells." (Preface, 1976, p.90).

We start from the assumption that the concept of underdevelopment mystifies the process by which capitalism gives rise to forms of development which are very uneven, both within social formations and internationally. [8] It does not seem logical to assume that the world capitalist system needs to depress the growth and expansion of its own structures by imposing blockages on the development of capitalism in some of its constituent social formations. It seems more logical to assume that capitalist development in a social formation is determined both by specific features of that social formation itself, and also by the position of the social formation in the world capitalist system and the international division of labour. [9] In other words, there is development under capitalism, but the forms it takes are variable.

Our understanding of the nature of these various forms of capitalist development will depend on distinguishing between two relevant totalities; social formation is a totality within capitalist development, but it is a totality which is subordinated to a greater totality, that of the world capitalist system. The nature of the transformation process is determined by the internal contradictions of particular social formations, and their class practices, but these contradictions are also constituent parts of the world capitalist system. The organisation of social relations of production, and the formation of class practices are not only governed and structured by the internal dynamic of the social formation, but also by the motions of capitalist development on a global scale.

1. Domination by the capitalist mode of production

In order to account for the different forms taken by uneven development, it is essential to be able to recognise, in particular cases, when the dominant mode of production has become capitalist. There are obviously a number of problems here, but in general, after the necessary initial stage of primitive capital accumulation, the following processes have to be identified.

(i) Commodity production must have become the general form of production. Commodity production can exist under pre-capitalist modes of production, so one of the main processes which must be analysed is the extent to which commodity production has become generalised.

(ii) The spheres of reproduction must be established for the expanded reproduction of capital. To understand this we make use of Marx's distinction between formal and real subsumption of labour under capital. By the formal subsumption of labour under capital, Marx meant a situation in which capital takes over an existing labour process, ushering in a situation which is changing but which is still only formally distinct from an earlier mode of production and one in which labour produces absolute, rather than relative surplus value. [10] Two developments then follow: firstly, the relationship between producers and owners becomes reduced to a purely economic relationship of supremacy and subordination, as opposed to the more complex relationships of small scale pre-capitalist production, and secondly, labour becomes used far more continuously and intensively. At this stage, Marx envisaged an increasing diversity developing in opportunities for earning one's living, but small capitalists would still be a little different from their workers in terms of their education and activities.

The real subsumption of labour under capital is explained by Marx as follows:

> "capitalist production proper begins only when capital sums of a certain magnitude have directly taken over control of production either because merchant turns into an industrial capitalist, or because larger industrial capitalists have established themselves on the basis of the formal subsumption."
> (Marx, 1976, Appendix p.1027)

At this stage, the nature of the labour process is transformed. Relative surplus value is produced, the social forces of production of labour are developed, and science and technology are directly applied to large scale production. The increased scale of production, and the creation of "surplus population" calls new branches of industry into being, and in these new branches, small scale production starts again with the formal subsumption of labour under capital, in a continuous process. The logic of capitalism means that development proceeds to the stage of production for the sake of production, but also that different stages of production can coexist. [11] It is this diversity of stages of production which can be confusing in identifying the development of the capitalist mode of production in a given social formation.

(iii) The transformation of small direct producers into wage earners (the process of proletarianisation) must be completed on a large scale, thus subsuming the so-called peasant producers under capital in different ways (Bradby, 1975; Banaji, 1977; Bernstein, 1977). This is a crucial process for the organisation of capitalist production relations and the social appropriation of the means of production.

(iv) Capitalism is characterised by an effective separation of the economic from the political, and by the state taking a distinctive role in the development of relationships between modes of production. As capitalist production develops, this role becomes vital for the generation of class practices and for the resolution of contradictions. This occurred in the original emergence of "western capitalism", but it is also occurring today in different ways, with the incorporation of social formations into the existing world capitalist system. Different states play different roles in mediating between the interests of fractions of indigenous capital and the interests of international capital, and we will return to this point below.

In any discussion of the general conditions which must be fulfilled for the capitalist mode of production to dominate other modes in a social formation, there are certain general qualifications which need to be made. Firstly, as we have argued above, the transformation of social formations in the original development of capitalism differed considerably from more recent developments. Secondly, the introduction and development by international capital of generalised commodity production in developing social formations, leads to the breakdown of local self-sufficiency in the basic units of production, and so to enforced production for both domestic and world markets. This is the argument that imperialism pioneers the development of capitalism (Warren, 1980). Thirdly, the expanded reproduction of capital takes different forms today from those which appeared in the initial growth of capitalism. Although surplus is produced, and this surplus can generate a degree of development, the structures of integration which have now developed between emergent capitalist social formations and the world capitalist system, maintain exploitative production relations. This perpetuates the uneven development of capitalism not only within social formations, but also within the world capitalist system. The trend is towards the concentration of capital into fewer centres than before, because the logic of capitalist development demands a restructuring of the capitalist division of labour internationally. In other words, although the emergent capitalist social formations are developing, albeit unevenly, they are being incorporated into an existing world capitalist system on more or less disadvantageous terms.

The problems of how to recognise the domination of the capitalist mode of production and the emergence of a capitalist state in a given social formation are inevitably extremely complex. Here we have tried to indicate in a very general way the main criteria which will be needed, but in order to understand the complexity of transformation, detailed questions need to be posed, which raise considerable theoretical and methodological problems about the state and class in emerging capitalist social formations. [12]

2. Class struggle and the state

Class struggles can only be studied through analysis of the organisational structures of the state, and the ways in which the functions of particular state apparatuses affect the matrix of various class practices. Such studies must necessarily be specific to particular social formations. Since states only exist in social formations, only qualified generalisations can be made about them at an abstract level. There are, nevertheless, some key general characteristics which need to be specified in order to identify particular capitalist states. It can be stated in general that the state is the arena in which heterogeneous class practices come together in their crystallised forms, and that it is the historical role of the state to maintain existing structures of relationships by contributing to the reproduction of the conditions of existence of the capitalist mode of production. The nature of each capitalist state is determined by particular configurations of social classes and by struggles for access to decision-making processes in the distribution of power.

State power can be controlled by different fractions of the dominant social classes, under different conjunctures of domination and subordination. Subordinate classes can be represented within the state structure as long as their demands do not extend to the effective sharing of political power. A capitalist state is one in which the capitalist class dominates state power, but the state is unlikely to serve the interests of the capitalist class exclusively. We need to understand the many variations in the struggles between classes and fractions of classes to gain access to state power, and the degree to which state apparatuses become penetrated by the dominant capitalist interests. Separation of the political from the economic is effected when the subordinate classes receive increasing economic benefits, but lack political dominance. Varied class struggles can develop, but it is only in capitalist systems that class struggles are increasingly presented and resolved as if they were economic issues.

We take the view that in every case the state must attempt to control the transformation process, and the variable ways in which the state can be organised contributes to the determination of the transformation process. Many of the differences between the areas of the world which have been labelled semi-periphery and the rest of the Third World need to be analysed at the level of variations in state organisations, state intervention in the economy and class struggles which are generated. The ownership and control of the means of production in any given social formation are determined by the ways in which production relations are organised, and this means that the class struggles which develop become regulated by the functioning of the apparatuses of the state. It follows from this that the political power which is exercised within the framework of the relationships of domination and subordination, becomes confined to the nature and organisation of these apparatuses. The state then functions to maintain the existing structures of relationships and to reproduce or to inhibit the development of the conditions of existence of the dominant mode of production, but in different periods of transformation, these processes can take different forms. The extent to which class practices can play significant roles in this transformation process depends on how these state apparatuses are organised, while the degree to which a social formation can be penetrated by centralised power relationships is a function of the level and the nature of the development which has taken place in that social formation. This is not only a political and ideological problem, but also an economic one.

Every state structure, during the transformation of the social system, must play a distinct and important role in trying to secure the effective functioning of its apparatuses. This means that every aspect of life in that social formation can be penetrated and that the effects of class struggle can be monitored in accordance with what appear to be the requirements for maintaining and reproducing the dominant social relations of production. The role of the state in the transformation process is, therefore, relatively autonomous at times. In other words, in the struggle to provide the necessary conditions for the reproduction of the emergent production relations, the state is not simply the tool of a ruling class, but can act independently of any structural constraints imposed by the configuration of dominant and subordinate classes. [13] The state can act relatively autonomously in that it can transcend specific, dominant class interests and their fractional forms. Different states are thus able to intervene directly or indirectly in the promotion of capitalist development in quite different ways, and with different consequences. Since this is so, we can make the generalisation that the nature of the state's role in the transformation is determined by class struggle, but only with the qualification that class struggles can only exist, and can only be studied in the specific forms which they take in historically determined social formations.

The importance of this point is not that it is new, but that it is easier said than done. Too often the specificity of states and classes is forgotten in a welter of generalisations on peripheral capitalism which stop short of clarifying the differences between actual cases. Even in the most general terms, capitalist states must be seen as developing in variable circumstances. The development of class struggles, for example, is necessarily affected by the extent to which the state apparatuses can control the transformation process. We cannot generalise, however, without knowledge of each case, about what struggles will actually develop in each social formation, since the nature of class struggles will be determined by particular matrices of internal and external contradictions.

It is this point which makes it essential to look at the ways in which the world capitalist system is maintained and reproduced. If we are to understand the uneven transformation of the social relations of production in a social formation, we need to identify the unique historical character of the social formation, but we must also account for the place of the social formation in question within the world capitalist system. In the next section we consider how the world capitalist system operates.

THE MAINTENANCE AND THE REPRODUCTION OF THE WORLD CAPITALIST SYSTEM

Turkey, like any other social formation today, is part of the world capitalist system. It is essential, therefore, to understand Turkey's position within the international division of labour, since it is this division of labour which forms the basis of the world capitalist system and determines its very nature. Just as it would be a mistake to look at Turkey in isolation from the world system, it would be a mistake to examine the world capitalist system as if it had a separate existence from the social formations which comprise it. The world capitalist system does exist as a system with interconnected parts, but it does not exist independently of these parts. [14] When we talk of the world capitalist system we are referring to the social, political and economic activities of a multiplicity of social formations which differ from each other in a multiplicity of ways. [15]

If it had been the case that the world capitalist system simply determined the ways in which its constituent social formations could develop, our understanding of the complexities of the development of world capitalism would have been tremendously simplified. In practice, however, the world capitalist system is not uniform, and the internationalisation of capital alone does not explain the determination of the levels of development of a social formation. There are serious contradictions which plague the world capitalist system, but since the system is only the sum of its parts, these contradictions must be rooted in its parts. That is, the contradictions of the world capitalist system, which generate, for example, over-production, surplus labour or energy crises are translated into national forms in different social formations. There are different forms of over-production and then only in some areas of economic activity and at particular periods, in particular economies; different forms of surplus labour emerge in different areas in particular combinations of circumstances; there have been different energy crises with different social, economic and political consequences, and so on. As these general contradictions of world capitalism take specific forms, they become part of the whole set of contradictions specific to each social formation. Ultimately it is this combination of internal contradictions which determines the relationship between any given social formation and the world capitalist system.

The world capitalist system has a number of additional characteristics which serve to complicate its nature and the way in which its social formations can develop. For example, the world capitalist system incorporates not only capitalist nations but also, in broad terms, socialist ones as well. The socialist nations obviously differ in certain respects from capitalist ones, but they share a set of rules which provide a common framework for political and economic action at the international level.

Another complication is the persistent conflict between capitalist nations over the patterns of specialisation in production, control of world markets and the degree of influence each nation-state can exert over the activities of multinational companies. In attempting to understand the immense complexity of the world capitalist system as a whole, it is necessary to simplify and to generalise, but it is essential not to forget that the operation of the capitalist system on any level is far from harmonious. The contradictions of the system are very real, and they develop into a variety of struggles which take a variety of forms. In general, these highly complex relationships between capitalist nations are due to the uneven development of capitalism within national economies, as well as internationally.

A related characteristic of the world capitalist system arises from the volatility of situations where attempts are made to curtail the expansion of capitalism by various revolutionary movements. These movements generally have some degree of support from socialist nations which, while operating within the parameters of the world capitalist system, are also set on changing the domination of capitalism. [16] In the battle between capitalism and socialism, much of the Third World has come to provide convenient battlegrounds for competing interests. Where the nations are sources of raw materials or where they occupy strategic political or military positions, they are particularly liable to become directly or indirectly caught up in the struggles of the dominant nations. [17]

The world capitalist system is inherently unstable, not only because of these threats from socialism, but because the strategies which are essential for the survival of the interdependent parts of the system, are also sources of instability. The need for an increasingly specialised international division of labour leads to an increasingly uneven distribution of political power. In addition, the expansion of capitalism means that as much of the world as possible needs to be brought as swiftly as possible into the sphere of capitalist relations, but this rapid expansion of capitalist relations continuously destabilises the international division of labour, and leads to changes in the distribution of political power within the world system.

This changing balance of power is often less apparent than it might be, because political struggles are frequently presented as economic necessities. In 1973, for example, oil prices were raised by the oil-producing countries, which later formed an organisation to defend their interests (OPEC), with drastic and irreversible consequences for the reproduction of capitalism on a world scale. This was a political act in that the aim of the oil-producing countries was to break down the advanced capitalist countries' monopoly of political power by using oil prices to destabilise the system, and they have been, by all accounts, very successful in achieving their objective. It was not, as some people thought, action against the major oil companies, but against the major consumers of oil who could between them decide on levels of development in the world system, by controlling the allocation of resources. The events of 1973 highlighted the vulnerability of the world system, and showed that access to decision-making could be achieved by using economic means to achieve political ends.

This example indicates another characteristic of the world system which works against the maintenance of an equilibrium. This is the necessity of maintaining relationships between former colonies and dependencies and their former imperial powers. What has come to be called the North-South relationship, with its goal of a New International Economic Order which will steadily erode the inequalities between rich and poor nations, is simply a label which conceals the increasing integration of the newly industrialising and more recently independent countries, on generally disadvantageous terms, into the world capitalist system. [18] It is one thing to aspire to be politically independent, and quite another to have to compete within an interdependent world system where political sovereignty and national economic independence are subordinate to the accumulation of capital on a world scale. When these international contradictions become apparent, they help to sharpen the contradictions which are internal to social formations. Increasing demands for improved conditions of existence, and popular resistance to foreign interests, leave the leaders of new nations with a limited range of options in the face of the reality of their place in the world capitalist system. What tends to characterise successful political leadership in these circumstances is the ability to sell policies geared to the development of a national capitalist class as nationalist policies aimed at improving the living conditions of the masses.

Since the world capitalist system is not a unified or harmonious system, its constituent parts do not act simply together either for or against each other. It is important for the dominant powers, however, that the needs of the system as a whole are defined in a way which is consistent with their interests. There are now a number of international agencies, such as the IMF, the World Bank and the GATT which serve as regulatory agencies for the expansion of capitalism on a world scale, and for the balance of the system as a whole. [19] These agencies provide development guidelines for the late-industrialising social formations, but they were set up originally to co-ordinate and to regulate general developments within the international division of labour. These two tasks are clearly linked together in such a way that financial and technical aid programmes tend to be shaped and restricted by the more general political need for a stable international division of labour. These international agencies exist in order to promote capitalist growth and development as widely as possible, but new developments can only take place within the existing framework of relationships of the world capitalist system. This promotion of capitalist development takes the form of integrating ever more fully into the system the late-industrialising social formations, through "aid" to the Third World, but the agencies also boost capitalist growth in established parts of the system, such as Britain in the 1970s, where growth may be flagging.

The foundations of the present world capitalist system, and its regulatory institutions were laid by the Bretton-Woods Agreement of 1944. The first initiative for an agreement had come from the United States in 1941, and the British were the first to be approached. Negotiations between these two countries from 1941-44 culminated in the establishment of the International Monetary Fund in 1945, and the International Bank for Reconstruction and Development (World Bank) also in that year. The General Agreement on Tariffs and Trade (GATT) was set up in 1948, in place of a proposed International Trade Organisation, a much more ambitious agency, which failed to get off the ground.

The ideas which inspired the creation of a system to regulate the development of capitalism on a world scale were political in essence and economic in form. The depression of the inter-war years and the crisis of Second World War prompted the recognition of a world economic order which could be regulated by international agencies, so that the recurrent crises of capitalism could be monitored by its member states. The famous American memorandum of 1942 formulated the guidelines for future action:

> 1. An international organisation for maintenance and exchange stability and to deal with balance of payments problems,
>
> 2. An international organisation to deal with long-term international investment,
>
> 3. An international agreement on primary commodity price control,
>
> 4. International measures for the reduction of trade barriers,
>
> 5. The international organisation of relief and reconstruction, and
>
> 6. International measures to maintain full employment.
> (Penrose, 1953, pp.39-40)

These guidelines formed the foundation of the world capitalist system and, except for No.6, are still consistent with the measures recommended today by the IMF and the World Bank. Policies based on these guidelines are thought appropriate for those nations which are deemed to be unsound in their capitalist development, those nations which are still not fully integrated into the world capitalist system, or those which, for political or ideological reasons, do not consider themselves capitalist but have to ask for international economic assistance.

In recent years the IMF has undertaken a number of assignments to restructure the economies of member states. Since the IMF is the regulating agency for the international monetary system, international money markets are greatly influenced by the stance vis-a-vis member states that the IMF takes. The Fund, as it is sometimes called was established with the aim of eventually replacing the gold standard, and thus avoiding the recurrence of the chaos of the thirties which led to the near collapse of the system. The IMF has had a chequered history, but in recent years it has emerged as the leading international economic regulatory institution, particularly since the oil crisis of 1973 when the subsequent world recession brought home the reality of how much direction is required to maintain any stability.

The Fund usually recommends policy packages which are geared to the abolition of trade barriers and inward-looking economic strategies. Its interventions so far have been designed to stimulate the growth of national economies by strict monetary policies. These policies aim to reduce rates of inflation by the control of money supply, to privatise and concentrate capital and to reallocate scarce resources from uncompetitive to competitive industries; competitiveness being determined, a) by the patterns of specialisation in the world markets, b) by the highest return on investment, c) by changes in the composition of capital, d) by the setting of "realistic" price levels and e) by the improvements in the quality control.

Obviously all these measures have political implications, and to present them merely as guidelines for economic recovery based on value-free economic analysis is seriously misleading. The assumption that there is no alternative but to follow these policy recommendations is in itself a political decision because it reflects a choice between available alternatives, although admittedly some strategies are more realistic than others. The IMF policy packages are offered to states which are in difficulty on the assumption that they will continue to promote capitalist development and to fulfill a designated role in the maintenance of the world capitalist system.

Since socialist states are also incorporated into the world capitalist system, they are encouraged to borrow in international money markets and to trade in the world commodity markets. The IMF will also offer them assistance, for example, with servicing of debts in times of economic difficulty. [20]

Even within this framework, such policy recommendations give rise to enormous political and social divisions in countries where they are being implemented. Usually the process of restructuring the economy and of tying it more firmly to the world markets, raises all sorts of political issues that have repercussions far beyond the scope of change that monetarist policies allow for. The IMF has often been accused of recommending policy measures which are out of step with existing social and political patterns of national development. The Fund and its policy makers are accused of insensitivity and even total disregard for national feelings. Probably such accusations have some substance, but they do not recognise the paramount preoccupations of the Fund, namely to maintain the stability and the reproduction of the world capitalist system, to which national sentiments must be subordinated. We would argue that the Fund, together with the World Bank, has so far achieved tighter integration of the world system at the expense of many national sacred cows, thus contributing to the resolution of some of the important contradictions in the capitalist development of member states. This achievement has not simply been due to imposition of policies by international agencies on unwilling states. With the development of capitalism on the world scale, internal class conflicts have sharpened, while ruling classes have proved unable to solve national economic problems within existing political structures. This has led to a variety of situations in which dominant interest groups have welcomed intervention by international capital and its agencies.

The single-minded determination to resolve the contradictions of capitalism which is characteristic of IMF policy, has been effectively demonstrated on several occasions. In 1976, for example, Britain, which was after all one of the two initial originators of the present framework for regulating the world capitalist system, had to go to the IMF to negotiate a massive loan. The conditions imposed on Britain by the Fund were, with minor differences, very similar to those imposed on, for example, Turkey, Peru, Sri Lanka, India and more recently on Mexico and Brazil. Despite Britain's historic position in the world capitalist system, it has received essentially the same treatment as other less-developed countries. The ramifications of this event have been so profound for Britain that the British social formation is at present undergoing a transformation that will tie it more closely to the world economy than ever before. What this means is that British capital accumulation will continue primarily in world markets but not necessarily in Britain. The British economy will produce its allocated share, which is determined by the mechanism of the world markets, and become a net importer of goods produced by other economies operating on the same basis. The concept of "national economy" is simply becoming irrelevant in accounting for the operation of the world capitalist system, and is in the process of being scrapped from the textbooks. The political effects of IMF policies on British society can be seen in a discernible trend towards increased privatisation of property and the concentration of capital into fewer hands. Mass unemployment is rapidly becoming normal as the organic composition of capital is drastically and persistently altered at the expense of labour-intensive technology. These economic trends are accompanied by political trends toward the establishment of an increasingly right wing political platform. This is again no different from many other examples in Newly Industrialising Countries of which Turkey is one. We are not suggesting that these social formations have a state structure that is comparable with Britain's, but that they all share economic strategies which can only be realised through unrepresentative politics.

It seems quite clear that there is only one world system and that its parts are a network of interdependent member states and their interrelationships. Attempts to distinguish between its North and South, or to classify parts of it as Third World are misleading. However disadvantaged individual states may be in the international division of labour, they are all incorporated into the world capitalist system, and all parts of the system are characterised by uneven development. There is undoubtedly a great deal of direct and indirect exploitation, but there is also development under capitalism, however uneven this may be. Kay very usefully makes the point that where capitalism is undeveloped, the

It is equally problematic to see how such an economy can successfully withdraw from the world capitalist system to some state of national self sufficiency. Far from promoting capitalist development, this inward-looking economic policy actively blocks the integration of developing social formations into the world capitalist system and yet it is only through this integration that the development of capitalism can be promoted. The blocking of capitalist development, however, does not mean that this hybrid policy leads to the realisation of socialist aims; Tanzania is a case in point here. [23] The attraction of the Soviet model lies not in its economic efficacy but in its ideological appeal to the nationalist and anti-imperialist sentiments of the people of the newly independent nations. Social scientists have frequently been caught up in these contradictions without always fully recognising them, as the debates on Tanzanian development, for example, show.

As a result of the popularity of this contradictory model of economic development, many of the "underdeveloped" nations of the world have followed similar development strategies. In particular they have adopted an inward-looking model of economic development with a strong protectionist trade regime and an import-substitution programme of industrialisation. This strategy of economic development generally requires an authoritarian political framework, although formal appearances vary, in order to implement it. In spite of its popularity, this strategy has proved effective only for limited periods, as once domestic markets cease to expand, inward-looking growth gives way to stagnation, and expansion requires looking for export markets. There are now signs that import-substitution programmes are being displaced from the agenda of strategies for development by much more realistic models of economic development. These are outward-looking and emphasise export-promotion and thus are in correspondence with the prerequisites for the development of capitalist social formations. By reducing the contradictions of capitalist development these development strategies may eventually give rise to much more humane and politically representative capitalist systems. In the meantime, while the transformation of economic strategies is taking place, the political spectrum will still be dominated by right wing politics, as we can see in Turkey, Argentina, Brazil, India etc. There is still, however, the possibility, if not the probability of the eventual development of representative politics once the contradictions between the fractions of the dominant classes are resolved and a stable state structure reflecting the unity of political and economic power is established.

It is essential to understand the importance for the future balance of world power of the changes in economic development strategies that are now being promoted and adopted. The New International Economic Order provides a set of political and economic relationships within which the restructuring of the world capitalist system is well under way. Although this is a slow and lengthy process, existing structures of domination and subordination in the system are being significantly altered, and it would be naive to suppose that such developments (as the accretion of economic power to OPEC has shown) could be free of conflicts. Whatever form these conflicts may take, countries like Brazil, Mexico, Korea, Turkey, Indonesia, Greece, Taiwan, Singapore and others are likely to become rapidly more competitive, provided that they can maintain relatively stable political regimes.

The long-term effects of the consequent shifts in the power structure of the world system, should not, however, be looked for in terms of traditional concepts such as nation, state, sovereignty, independence,

social formations in question have been insufficiently exploited (Kay, 1975).

If the view of capitalism as a central parasite preying on its own periphery is reconsidered, we can see that it is much more probable that some development takes place in all parts of the world system. [21] Imperialism has taken the form of the continuous internationalisation of capital, creating development, albeit uneven development as Third World countries become incorporated into the world capitalist system. Rather than pursuing a myth of generalised underdevelopment, or a transition to peripheral capitalism, we need to look at the real opportunities which the transfer of technology has given to the late-industrialising economies to compete in expanding world markets.

If we use the distinction between mode of production and social formation, it is possible to see that within the general development of capitalism, the transfer of technology can be a different process in different social formations. Since not all economies are affected in the same way or to the same degree, there is very considerable differentiation within the Third World. As capital has become increasingly international, the widespread availability of advanced technology and know-how has been a major factor in turning national economies away from inward-looking development strategies towards outward-looking policies. Thus the transfer of technology can contribute to the promotion of sustained growth if it is associated with an outward-looking, export-oriented economy. Higher levels of growth and increases in standards of living have also been achieved where belated recognition has been given to the growth potential of the agricultural sector.

The transfer of technology has taken on a particular significance for the economic competitiveness of the Third World over the last two decades, as a result of increasing specialisation in the production processes of the world economy. For example, in strictly economic terms, Britain does not need its own steel industry (with the possible exception of some specialised steel capability), or its own textile industries, if steel and textiles can be imported from, e.g. Southeast Asia, where the same products are produced much more cheaply, efficiently and with better quality control. Where technology is readily available, newly industrialising countries can obtain this advanced knowledge and production techniques, and with the help of relatively cheap labour can produce them much more competitively than Britain can. The possible conflicts of interests that arise here are outside the strictly economic sphere.

Until relatively recently, however, there was little emphasis in the later industrialising economies on manufacturing for the export market, or on investment in capital intensive agriculture in order to achieve sustained growth. The development priorities which were generally characteristic of the supposedly underdeveloped social formations came from adopting a model of development which was a curious mixture of classical liberal economics and policies derived from the Soviet experience of industrialisation for the domestic market. [22] The situations identified as underdevelopment and peripheral capitalism, then can be in part attributed to the choice of this model of development.

The Soviet model of economic development, which formed the basis for the development of this curious mixture, differed significantly from a capitalist model. This put the economies of the Third World in a difficult situation. When an economy is following the capitalist path, it is contradictory to expect it to achieve capitalist development through policies modelled, even indirectly, on a socialist economy and politics.

etc. At the moment, the units of the world system are more or less discrete economic and political entities, for which such terms are appropriate. Present trends in the development of the international division of labour, however, seem to be moving towards an international system of sectoral differentiation, and this trend seems to be inevitable if capitalism is to survive. This will mean an increasing tendency for national economies to become so specialised that they will only be able to function together with equally specialised but complementary economies elsewhere. Our traditional concepts will not allow us to indicate the nature of this new reality and we shall need to develop new ones.

The degree of interdependence of the parts of the world capitalist system is the source of its strength, but at the same time, its critical weakness. [24] The continuous restructuring which the survival of the system seems likely to require, cannot be achieved without the danger of provoking severe crises, and war. We would argue that the New International Economic Order has been put forward as one possible means for averting the inevitability of widespread conflict. It is paradoxical that while the dangers are widely recognised, no dominant nation seems to be making a serious attempt to prevent a nuclear solution to the contradictions of the world capitalist system. The armaments industries are booming, the arms race is intensifying, and the possibility of total annihilation comes ever closer. In these circumstances, any long-term planning for the future becomes either a refusal to face reality, or a form of depression.

This paradox is of course, understandable when the role of the armaments industries both in national economies and in international trade is analysed. The importance to domestic production and the volume of trade created is so vast that the whole edifice of world capitalism could collapse without it. This situation has come about not from simple economic necessity, but from interrelated expediencies. Resources could be reallocated for investment in more productive sectors, with appropriate readjustments throughout the system. The development of world capitalism, however, does not seem to be taking this path. Once nuclear weapons have been developed, the chances of limiting all wars to non-nuclear combat becomes intensely problematic. Unless some means can be found to reduce and eventually to eliminate nuclear arsenals, the continued restructuring of world capitalism will take place in the shadow of Armageddon.

Our purpose in making this gloomy appraisal of our future is to emphasise that capitalism is a truly international system, and it can only survive if it can be continuously reproduced on a world scale in an expanded form. Any forces which shackle these processes of reproduction will be countered by fierce resistance, and when contradictions cannot be resolved peacefully, war is the only alternative. Nuclear war is not an economic necessity of the system, but a consequence of the accumulation of political choices which have been allowed to gain their own momentum.

The following chapters in this book all focus on Turkey, but since the capitalist system now exists as an international whole, it does not make sense to treat its constituent social formations in isolation. Turkey, like all the other social formations that comprise world capitalism, makes its own distinctive contribution to the reproduction of the world system, and simultaneously is determined by the specific nature of the contradictions of the world system; yet our knowledge of these processes is relatively limited. The case of Turkey is, therefore, of concern not only to Turkish specialists, but to all those interested in understanding the operation and reproduction of the world capitalist system. Scholars cannot prevent the nuclear lunacy of our time, nor the inhumanity which citizens experience at the hands of their own states, but they can produce knowledge and they can make that knowledge more widely available. Through analysis of different social formations perhaps we can become more familiar with different parts of the world system, their internal operation and the ways in which they work together, in order to provide a better basis for sane participation, regulation and control, at all levels of society.

NOTES

1. An earlier version of this paper, "The Articulation of Modes of Production, Class Struggles and Labour Migration", was written in collaboration with J. Bujra and was presented at the British Sociological Association Conference, Aberystwyth, 1981. Bujra's contribution is being reworked for inclusion in a forthcoming book on Kenya, edited by J. Bujra and M. Cowan. We are grateful to Bujra for many comments and criticisms, but responsibility for the views we have expressed here remains with us.

2. Although parts of the world may be termed socialist, these are also interdependent parts of the world capitalist system.

3. For a brief examination of this view in terms of the levels of analysis required to achieve an understanding of the specificity of social formations see, Ramazanoglu, 1980.

4. The discussion of this issue in its most original form is the debate which took place in <u>Science and Society</u> in the 1950s; see Hilton, 1976. Similar issues are also discussed by Resnick and Wolff, 1979.

5. Some authors, notably Hindess and Hirst, have seen transformation as part of transitional conjunctures, but they have not clarified the links between transition and transformation. See Hindess and Hirst, 1975, pp.283-284; and critical reviews by Cook, 1977 and Taylor (1975, 1976).

6. For a summary of these arguments, Wolpe, 1980.

7. Some of the contributions to the debate on the transformation of pre-capitalist modes of production in India illustrate this point particularly well. These contributions can be found in Rudra et.al., 1978. For a very good assessment of this debate see Harriss, 1980; also Alavi, 1975; Cleaver, 1976; McEachern, 1976.

8. Explanation of this point can be found in Philipps, 1977. The late Warren devotes his whole book to a critique of Lenin's thesis on imperialism and its adoption by post-war Marxists. Warren, 1980, especially pp.111-14.

9. This point can be made for any social formation.

10. Absolute surplus value is achieved by lengthening the working day

to increase productivity. Relative surplus value is achieved by technological inputs being used to increase productivity thus decreasing the proportion of the working day in which workers work for their own subsistence.

11. See Marx, 1976, p.1025ff.

12. Brenner, 1977, in his intervention in the debate points out that class struggle determines the relationships between different modes of production, but does not pursue his argument to include the role of the state. The relations between class and state need further elaboration in Marxist theory. See Poulantzas (1973, 1975, 1978); Hindess and Hirst, 1975; Hindess, 1977; Hunt (1977, 1980); Wright, 1978; Urry, 1981; Jessop (1982, 1983); Clark, 1983; Held, et.al., 1983.

13. For example, Trimberger, 1977, examines the relatively autonomous role of the state during the transformation of the Japanese social formation.

14. There is a wide ranging literature which examines the world capitalist system. There seems to be general agreement in the field that capital accumulation is taking place at an international level and that this accumulation determines that nature of the international division of labour. Views diverge, however, on the positions of the constituent social formations of the world capitalist system itself. The disagreement is over whether the system has an existence as such, in that it can act independently of its parts or whether it is no more than its sum of its parts. Here we subscribe to the latter view.

15. There are a number of works which deal with these complex issues, but it is not always clear how their authors view the world system. The series of Sage Publications, <u>Political Economy of the World System Annuals</u>, published under the general editorship of Immanuel Wallerstein makes the "world system" view very clear. This series takes the view that the world system is more than the sum of its parts and has its own discernible identity, and as a result of this view it suffers from the serious shortcomings of such a position. Wallerstein and his Fernand Braudel Institute at Binghampton, N.Y., have become the main source which propagates these views and, in recent works in the Sage series has confronted some of the criticisms made of dependency theory.

    Amin (1974, 1976, 1977, 1980) has also made very similar arguments to the ones advocated by Frank and Wallerstein. Authors such as Emmanuel, 1972 and Arrighi, 1978 have also contributed to the debate within the methodological, theoretical and political framework of of the "world system" approach.

    The assumptions on which dependency theory rests, however, while increasingly elaborated, remain fundamentally unchanged. A good discussion of Frank-Amin-Wallerstein link can be found in Worsley, 1980.

16. There is an influential body of literature, primarily by Soviet scholars, which advocates the demise of capitalism as the main goal of international revolutionary movements. These authors see the

newly independent and developing nations of the world as being areas where "the non-capitalist way of development" can be successfully implemented. They see these nations as areas where the expansion world capitalism can be effectively curtailed, and socio-economic transformation can be carried out within the framework of "socialist orientation". "The non-capitalist way of development" indicates the socio-economic aspects of anti-imperialist revolutions. The main justification for promoting non-capitalist development being the inability of world capitalism to solve the problems of newly independent nations of the world, and the growing influence of socialist countries as viable alternative modes of development. These scholars, therefore, see these emerging countries as necessarily having to make an "objective" decision to adopt the "socialist orientation" since this is the only way in which they can achieve national liberation and economic development. The leading social classes in the process of "socialist orientation" are national bourgeois elements and the radical petty-bourgeoisie which are untainted by imperialist and metropolitan bourgeois interests, and are, therefore, able to lead an anti-imperialist movement.

For further reading on Soviet theses on development see, Solodovnikov and Bogoslovsky, 1977; Andreyev, 1977; Ulyanovsky, 1978. For an analysis of these theories see, Clarkson, 1979.

Although the applications of these theses differ in some respects from those of dependency theories, both schools of thought owe a debt to Lenin, and share common assumptions on the nature of imperialism and the structure of the world capitalist system.

These theses are theoretically and politically so defective that it would require a long and detailed critical analysis to demonstrate their inadequacies fully. This is a task which we do not propose to undertake here.

17. Where, however, nations are not sources of raw materials and do not occupy strategic political or military positions, their existence can become marginal to the reproduction of the world capitalist system. These tend to be the poorest nations of the world which generate little surplus of their own, have little attraction to others, and which have little hope of benefitting from the international distribution of surplus.

18. The two reports by the Brandt Commission (1980 and 1983) are an indication of a major development in thinking in various centres of international capital. They indicate a move towards the need to increase the integration of the "South" into the world capitalist system through international aid and grant schemes which will stimulate economic development. The main aim of these reports are to draw the attention of the "North" to the need to decentralise the world capitalist system so that the reproduction of capitalist relations can be realised at different levels of the world capitalist system, thus strengthening further its conditions of existence. The Brandt Commission reports have come under fire from various sources. These criticisms are generally located within the "dependency" framework, see Graf, 1981; Hayter, 1981; Elson, 1982.

    Apart from the reports produced by the Brandt Commission itself, there are a number of other useful works on the subject of the North-South relationship. Works by Bhagwati (1977, 1984); Fishlow, Diaz-Alejandro, Fagen and Hansen, 1978; Spero, 1981; Anell and Nygren, 1980 and Anell, 1981; Frobel, Heinrichs and Kreye, 1981; Brett, 1983 are good sources to refer to for further reading. South magazine is also very useful source of factual information and comment on the current restructuring of the world economy and politics. Another useful source on the workings of the world capitalist system is Gauhar, 1983.

19. A succinct examination of the role of these international agencies in the development of the world economy since the war can be found in Block, 1977; Scammell, 1980; Brett, 1983; Armstrong, Glyn and Harrison, 1984. There are also many critics of these agencies, especially of the IMF and the World Bank. The United Nations Economic Commission for Latin America (ECLA) under Roul Prebisch has been one of the most vociferous opponents of the IMF and the World Bank. Hayter, 1971, takes a very critical view of the World Bank and other aid agencies. Payer, 1974, is devastatingly critical of the IMF policies, and she points out to the death of democracy in Third World countries as a result of the IMF interventions and their consequences. Latin America Bureau, 1983, has brought out a collection of critical theoretical and empirical essays dealing with the role that the IMF plays in the world capitalist system. Bird, 1982, on the other hand, presents a very detailed examination of the relationship which has developed between the LDCs and the IMF, without reaching any firm conclusions. Eringer, 1980, draws the public's attention to the crucial role of the overall framework of economic and political structures of the world system. The book falls far short of substantiating its claims, but, nevertheless, reveals the existence of these powerful groups to the general public.

20. There are two countries from the Socialist Block which are members of the IMF; Rumania since 1972 and Hungary since May 1982. Yugoslavia and Poland have also had substantial dealings with the Fund. The Fund played a key role during the Polish crisis of 1980-82, and continues to do so.

21. The late Bill Warren's intervention in the debates on the nature of underdevelopment has been most fruitful and honest, but not necessarily welcomed by everyone. For attacks on Warren see Emmanuel, 1974; McMichael, Petras and Rhodes, 1974; and also Lipietz, 1982. Warren has also been the subject of some very emotional attacks, see Ahmad, 1983.

22. These points have been amplified by Warren, 1979. These views are not original, but they emphasise a reality which had previously not been given sufficient emphasis. This curious mixture of the Soviet model of a planned economy with the assumptions of liberal economics has certainly been characteristic of the early stages of the development of Turkish capitalism. See also Ramazanoglu, Chapter 2, of this volume.

23. Mueller, 1980, brings out this contradiction in the Tanzanian case very clearly.

24. Similar thoughts must have played a key role in prompting Anne Krueger, 1972, in a speech made shortly before her appointment as

the Vice-President of the World Bank, to call for more direct intervention by the World Bank in the formulation of domestic economic policies.

REFERENCES

Ahmad, Aijaz (1983); "Imperialism and Progress" in R.H. Chilcote and D.L. Johnson (eds.), Theories of Development: Mode of Production or Dependency?, Sage, Beverly Hills

Alavi, Hamza (1975); "India and the Colonial Mode of Production", Socialist Register, Merlin, London

Alavi, Hamza (1982); "Structure of Peripheral Capitalism", in T. Shanin and H. Alavi (eds.), Introduction to the Sociology of Developing Societies, Macmillan, London

Althusser, Louis and Balibar, Etienne (1970); Reading Capital, New Left Books, London

Amin, Samir (1974); Accumulation on a World Scale: A Critique of the Theory of Underdevelopment, Vols.1-2, Monthly Review Press, New York

Amin, Samir (1976); Unequal Development, Harvester, Brighton

Amin, Samir (1977); Imperialism & Unequal Development, Harvester, Brighton

Amin, Samir (1980); Class and Nation, Historically and in the Current Crisis, Heinemann, London

Andreyev, I. (1977); The Non-Capitalist Way, Progress Publishers, Moscow

Anell, Lars and Nygren, Birgitta (1980); The Developing Countries and the World Economic Order, Frances Pinter, London

Anell, Lars (1981); Recession: The Western Economies and the Changing World Order, Frances Pinter, London

Armstrong, Philip, Glyn, Andrew and Harrison, John (1984); Capitalism Since World War II, Fontana, London

Arrighi, Giovanni (1978); The Geometry of Imperialism, New Left Books, London

Banaji, Jairus (1977); "Capitalist Domination and the Small Peasantry", Economic and Political Weekly, Special Number, August, pp.1375-1404

Baran, Paul A. (1957); The Political Economy of Growth, Monthly Review Press, New York

Baran, Paul A. and Sweezy, Paul M. (1968); Monopoly Capital, Pelican, Harmandsworth

Bernstein, Henry (1977); "Notes on Capital and Peasantry", Review of

African Political Economy, No.10, pp.60-73

Bhagwati, Jagdish (ed.) (1977); *The New International Economic Order: The North South Debate*, MIT Press, Cambridge, Mass.

Bhagwati, Jagdish and Ruggie, John (1984); *Pcwer, Passions and Purpose: Prospects for North-South Negotiations*, MIT Press, Cambridge, Mass.

Bird, Graham (1982): *The International Monetary System and the Less Developed Countries*, Macmillan, London

Block, Fred L. (1977); *The Origins of International Economic Disorder*, University of California Press, Berkeley

Bradby, Barbara (1975); "The Destruction of Natural Economy", *Economy and Society*, Vol.4, No:2, pp.127-61

Brandt Commission (1980); *North-South: A Programme for Survival*, Pan Books, London

Brandt Commission (1983); *Common Crisis, North-South: Cooperation for World Recovery*, Pan Books, London

Brenner, Robert (1977); "The Origins of Capitalist Development", *New Left Review*, No.104, pp.25-92

Brett, Edward E. (1983); *International Money and Crisis*, Heinemann/Westview, London and Boulder

Cleaver, Harry (1976); "The Internationalisation of Capital and the Mode of Production in Agriculture", *Economic and Political Weekly*, March 27, pp.a2-a16

Clark, Simon (1983); "State, Class Struggle, and the Reproduction of Capital", *Kapitalistate*, Nos.10/11, pp.113-30

Clarkson, Stephen (1979); *The Soviet Theory of Development: India and the Third World in Marxist-Leninist Scholarship*, Macmillan, London

Cook, Scott (1977); "Beyond the Formen: Towards a Revised Marxist Theory of Pre-Capitalist Formations and the Transition to Capitalism", *Journal of Peasant Studies*, Vol.4, No.4, pp.360-89

Elson, Diane (1982); "The Brandt Report: A Programme for Survival?", *Capital & Class*, No.16, pp.110-27

Emmanuel, Arghiri (1972); *Unequal Exchange: A Study of Imperialism of Trade*, New Left Books, London

Emmanuel, Arghiri (1974); "Myths of Development versus Myths of Underdevelopment", *New Left Review*, No.85, pp.61-82

Eringer, R. (1980); *The Global Manipulators*, Pentacle Books, Bristol

Fishlow, Albert; Diaz-Alejandro, Carlos F.; Fagen, Richard R.; Hansen, Roger D. (1978); *Rich and Poor Nations in the World Economy*, McGraw-

Hill, New York

Foster-Carter, Aidan (1978); "The Modes of Production Controversy", New Left Review, No.107, pp.47-77

Frank, Andre G. (1971); Capitalism and Underdevelopment in Latin America, Pelican, Harmandsworth

Frank, Andre G. (1969); Latin America: Underdevelopment or Revolution, Monthly Review Press, New York

Frank, Andre G. (1972); Lumpen-Bourgeoisie and Lumpen-Development, Monthly Review Press, New York

Frank, Andre G. (1975); On Capitalist Underdevelopment, Oxford University Press, Bombay

Frank, Andre G. (1978a); Dependent Accumulation and Underdevelopment, Macmillan, London

Frank, Andre G. (1978b); World Accumulation 1492-1789, Macmillan, London

Frank, Andre G. (1980); Crisis: In the World Economy, Heinemann, London

Frank, Andre G. (1981); Reflections on the World Economic Crisis, Hutchinson, London

Frobel, Folker, Heinrichs, Jurgen and Kreye, Otto (1980): The New International Division of Labour, Cambridge University Press, Cambridge

Gauhar, Altaf (ed.) (1983); South-South Strategy, Third World Publications, London

Graf, William D. (1981); "Anti-Brandt: A Critique of Northwestern Prescriptions for World Order", The Socialist Register, Merlin, London

Godelier, Maurice (1972); Rationality and Irrationality in Economics, New Left Books, London

Harriss, John (1980); Contemporary Marxist Analysis of the Agrarian Question in India, Working Paper, Madras Institute of Development Studies

Hayter, Teresa (1971); Aid as Imperialism, Pelican, Harmandsworth

Hayter, Teresa (1981); The Creation of World Poverty: An Alternative View to the Brandt Report, Pluto, London

Held, David Held, et.al. (1983); States and Societies, Martin Robertson/Open University, Oxford

Hilton, Rodney (ed.) (1976); From Feudalism to Capitalism, New Left Books, London

Hindess, Barry and Hirst, Paul Q. (1975); Pre-Capitalist Modes of Production, Routledge & Kegan Paul, London

Hindess, Barry (1977); "The Concept of Class in Marxist Theory and Marxist Politics" in J. Bloomfield, et.al., Class, Hegemony and Party, Lawrence & Wishart, London

Hunt, Alan (ed.) (1977); Class and Class Structure, Lawrence & Wishart, London

Hunt, Alan (ed.) (1980); Marxism and Democracy, Lawrence & Wishart, London

Jessop, Bob (1982); The Capitalist State, Martin Robertson, Oxford

Jessop, Bob (1983); "Accumulation Strategies, State Forms, and Hegemonic Projects", Kapitalistate, No.10-11, pp.80-111

Kay, Geoffrey (1975); Development and Underdevelopment: A Marxist Analysis, Macmillan, London

Krueger, Anne (1982); The Guardian, Wednesday, May 12, London

Lenin, V.I.O(1970); "Imperialism, The Highest Stage of Capitalism: A Popular Outline" in Selected Works, Vol.1, Progress Publishers, Moscow

Latin America Bureau, (1983); The Poverty Brokers: The IMF and Latin America, Latin American Bureau, London

Leys, Colin (1977); "Underdevelopment and Dependency: Critical Notes", Journal of Contemporary Asia, Vol.7, No.1, pp.92-107

Lipietz, Alain (1982); "Towards Global Fordism?" and "Marx and Rostow?", New Left Review, No.132, pp.33-58

Marx, Karl (1976); "Preface", Capital, Vol.1, Penguin, Harmandsworth

Marx, Karl (1976); Capital, Vol.1, Penguin, Harmandsworth

McEachern, Doug (1976); "The Mode of Production in India", Journal of Contemporary Asia, Vol.6, No.4, pp.444-58

McMichael, Philip; Petras, James and Rhodes, Robert (1974); "Imperialism and the Contradictions of Development", New Left Review, No.85, pp.83-104

Mueller, Suzanne D. (1980); "Retarded Capitalism in Tanzania", The Socialist Register, Merlin, London

Payer, Cheryl (1974); The Debt Trap: The IMF and the Third World, Pelican, Harmandsworth

Penrose, E.F. (1953); Economic Planning for Peace, Princeton University Press, Princeton

Phillipps, Anne (1977); "The Concept of Development", Review of African Political Economy, No.8, pp.7-20

Poulantzas, Nicos (1973); Political Power and Social Classes, New Left Books, London

Poulantzas, Nicos (1975); Classes in Contemporary Capitalism, New Left Books, London

Poulantzas, Nicos (1978); State, Power, Socialism; New Left Books, London

Ramazanoglu, Huseyin (1980); "After Dependency: A Contribution to the Debate", Journal of Area Studies, No.2, pp.22-25

Resnick, Stephen and Wolff, Richard (1979); "The Theory of Transitional Conjunctures and the Transition from Feudalism to Capitalism", Review of Radical Political Economy, Vol.2, No.3, pp.3-22

Rey, Pierre P. (1979); "Class Contradictions in Lineage Societies", Critique of Anthropology, No.13-14, pp.41-60

Rudra, Ashok, et.al. (1978); Studies in the Development of Capitalism in India, Vanguard Books, Lahore

Scammell, W.M. (1980); The International Economy Since 1945, Macmillan, London

Solodovnikov, V. and Bogoslovsky, V. (1977); Non-Capitalist Development: An Historical Way, Progress Publishers, Moscow

Spero, Joan E. (1981); The Politics of International Economic Relations, 2nd Edition, George Allen & Unwin, London

Taylor, John G. (1975, 1976); "Review Article: Pre-Capitalist Modes of Production", Critique of Anthropology, Part 1, No.4-5, pp.127-55, and Part 2, No.6, pp.56-68

Taylor, John G. (1979); From Modernisation to Modes of Production, Macmillan, London

Trimberger, Ellen K. (1977) "State, Power and Modes of Production: Implications of the Japanese Transition to Capitalism", The Insurgent Sociologist, Vol.7, No.2, pp.85-98

Ulyanovsky, R. (1978); National Liberation, Progress Publishers, Moscow

Urry, John (1981); The Anatomy of Capitalist Societies, Macmillan, London

Wallerstein, Immanuel (1974); The Modern World System, Academic Press, New York

Wallerstein, Immanuel (1980); The Modern World System II, Academic Press, New York

Wallerstein, Immanuel (1979); The Capitalist World Economy, Cambridge University Press, Cambridge

Wallerstein, Immanuel (1982); "Crisis as Transition" in S. Amin, et.al., Dynamics of Global Crisis, Macmillan, London

Warren, Bill (1979); "The Post-war Economic Experience of the Third World" in A Rothko Chapel Colloquim, Toward a New Strategy for Development, Pergamon, New York

Warren, Bill (1980); Imperialism: Pioneer of Capitalism, Verso, London

Wolpe, Harold (ed.) (1980); The Articulation of Modes of Production, Routledge & Kegan Paul, London

Wright, Erik O. (1978); Class, Crisis and the State, New Left Books, London

2 A political analysis of the emergence of Turkish capitalism, 1839-1950

HUSEYIN RAMAZANOGLU

In this chapter, I challenge the prevailing assumption that the Turkish social formation is an example of underdevelopment, dependent development or peripheral capitalism. Instead, I try to analyse the emergence of capitalism in Turkey as a specific form of the uneven development of capitalism. I reject the view that the development of world capitalism can usefully be understood through analysis of unequal exchange relationships, or that the social formations of world capitalism can usefully be labelled "centre", "periphery", or "semi-periphery". As I have argued elsewhere [2], each non-capitalist social formation, as it is incorporated into the world capitalist system, has to undergo a transformation in the course of which general processes of capitalist development have to be completed, but the completion of these general processes takes specific forms in each social formation. The position of any given social formation in the world capitalist system is determined by the specificity of the capitalist transformation process. There is a close relationship between international and indigenous capital, as part of the international division of labour, but in the last instance, the nature of this close relationship is determined by the internal contradictions of the social formation in question. Classifications of "centre", "periphery", or Third World are not useful, therefore, for understanding different stages of capitalist development. There is a world capitalist system, but each constituent social formation contributes to the reproduction of the world system differently, according to various conjunctures in its transformation processes. [3]

Here I take the view that Turkey has undergone a degree of capitalist transformation as a constituent part of the world capitalist system, and that capitalism creates uneven development; Turkey is, therefore, unevenly rather than under-developed. The study of the emergence of this uneven development of capitalism in Turkey, however, is the study of a particular case, and, therefore, necessitates an understanding of historical conjunctures which are specific to the Turkish totality. In this paper, therefore, I look empirically at key conjunctures of the Turkish social formation, and in particular at the development of class struggle and the formation of the capitalist state, in order to understand the rise to dominance of capitalist social relations of production.

This analysis is intended to provide a background to the origins of the current crisis of Turkish capitalism. Although the immediate causes of the crisis arise from the strains of opening Turkey's closed economy to world markets, the way the crisis has developed cannot be properly understood without a systematic examination of earlier periods. I start, therefore, by looking at the period from 1839-1923 when imperialist penetration of the Ottoman Empire created appropriate conditions of existence for the emergence of Turkish capitalism.

FROM TANZIMAT TO THE REPUBLIC: 1839-1923

There is general agreement among scholars studying the development of capitalism in Turkey that the critical period for the start of capital accumulation was the nineteenth century (Issawi, 1980). The process of accumulation was greatly facilitated by the free trade treaties of 1838-40 between the Ottomans, and Great Britain and other major European commercial powers. These agreements, which were the first trade concessions of the *Tanzimat* period (1839-76), resulted in hundreds of foreign merchants moving into what is now Turkey, and settling down to trade in every conceivable commodity. [4] This influx of foreign merchants had a devastating effect on the relatively inexperienced and undercapitalised native Ottoman merchants, who were then forced out of the market. During the 1840s and 1850s, trade between the Ottomans and England and France increased rapidly, with imports soon outstripped exports. This situation continued until the Crimean War, when the balance shifted drastically, and the Ottomans ended up with a considerable trade surplus. During this period the old craft industries were destroyed by foreign competition, and by the limitations imposed upon them by the powerful guilds.

The new Ottoman-produced manufactured goods were inferior in quality to European ones, and the Ottomans were still restricted by the impositions of the Capitulations first introduced in the sixteenth century, which gave trade advantages to the French and other European interests. It was quite remarkable in these circumstances that the Ottomans could develop a nascent industry of their own. The development of new industries took the form of the creation of entirely new manufacturing centres away from the influence of the guilds. New machines and know-how were imported from Europe to set up new factories, which produced clothing, cloth, headgear, rifles, shoes, cartridge belts and the like. The demand for these commodities came chiefly from state organisations and especially from the army. The quality of the products, however, was poor and failed to meet even domestic demand.

The new private factories which were being established throughout the nineteenth century were largely owned by foreign capitalists, with only a small percentage being owned by Ottomans. During the period of *Tanzimat* there was a constant flow of foreign capital into the Empire for setting up new units of production and factories to produce a diversity of commodities such as silk thread, silk clothing, carpets, flour, olive oil, tinned food, candles, paper, cotton yarn, cloth, rugs etc. Capitalist expansion also led to the exploitation of mineral resources, mainly in Anatolia. The "Mines Regulation" of 1861 ended the state monopoly of mines and allowed private ownership of land where mines were located, so that mining interests could be pursued and developed pri-

located, so that mining interests could be pursued and developed privately. The state was left with only those mines which were already on state land and which could be exploited profitably by the state. The state, however, lacked know-how, capital and technology for profitable mining so it leased state-owned mines to private companies. This effectively meant leasing to foreign capital, since there were no Ottoman private enterprises which could exploit mineral resources profitably. While the state treasury benefitted from taxes and royalty charges imposed on the operation of foreign enterprises, most of the mineral products, such as coal, manganese, lead, iron, silver and copper, were exported to be used in the production processes of the capitalist economies of Europe, thus critically restricting the development of heavy industry within the Empire itself.

The Tanzimat period provided the necessary legal and political framework not only for restructuring the organisation of the Ottoman state, but also for restructuring the economy as well. During this period, the integration of the Ottoman Empire into the world capitalist system began, and the emergence of a bourgeois class became possible (Kongar, 1976). It was this class that was to ensure that the framework for development laid down during the Tanzimat period was maintained. This new class was initially composed of foreigners and members of the ethnic minorities (Jews, Greeks, Armenians) who were the direct agents of foreign capitalist interests. During Tanzimat the ethnic minorities achieved equal social status with the Moslem Ottomans. Throughout the formation and the development of the Ottoman Empire, there had been discrimination between the people of the Empire. The Moslems, who were discouraged from taking part in trade were seen as the natural recruitment ground for military and administrative officials. Since Islam does not allow interest to be earned at the expense of others, Moslems were not encouraged to engage in corrupt earthly practices such as trade, instead they controlled those non-Moslems whose function in the Empire was to provide trading facilities for the Empire as a whole. The state was politically dominated by Moslems whose basic economic task was to collect taxes from the minorities who created the wealth; at the same time access to political power by the minorities was blocked. [5]

The Tanzimat policies forced a change in this system of relationships and broke up the Islamic structures of political domination by elevating the minorities to equal legal and political status with the Moslems. This resulted in a loss of power by the Moslems since they ceded political power to the minorities, but were not in a position to gain economic power from them. With their political advantage gone, the Moslem Ottomans were left in a weakened political and economic situation. This substantial loss of power by the Moslems marked the end of the period of the separation of economic and political power. Later on Ataturk and his followers interpreted the decline of the Ottoman Empire as being largely due to this separation of powers, and in the etatist period, which is discussed below, the repetition of such a separation was explicitly avoided.

As the Ottoman Empire developed a growing industrial sector, and international trade, the control and ownership of the means of production were firmly in the hands of foreign capital and its agents, the ethnic minorities. This situation allowed a bourgeois class to develop which was not moslem, but which, nevertheless, had the power to instigate and develop general processes of capitalist development. The emerging bourgeoisie was actively involved in setting the economic priorities of the Ottoman Empire with the help of European capital. As the power of the Ottoman Empire declined, the position of the minorities grew disproportionately stronger. Foreign relations (economic and diplomatic) became more and more important, and the only people who could handle the necessary correspondence and develop appropriate social, economic and political relationships, were members of the minorities. Their position in the economic, political and ideological levels became increasingly important because of changes in the class structure of the Ottoman social formation. While serving the interests of the Ottoman state, they also developed into committed allies of the European powers, and gained access to key positions at every level of the social structure. Although they did not occupy ministerial positions, nevertheless, by introducing new European methods of administration, professional education and technical training, these people established bridges between the traditionally insular Ottoman state, and the expansionist and highly industrialised European powers. Ultimately, the role played by the minorities led to the transformation of the existing social relations of production in such a way that they became the controllers of certain Ottoman revenues. [6] These they administered and collected and then devoted entirely to the service of debts incurred through wars and trade transactions. The Ottoman Bank and other Istanbul banks became the bastions of the financial power of the new non-Moslem bourgeoisie. The role of the state developed quite clearly as that of providing unqualified support for the reproduction of capitalist relations of production and the establishment of capitalism as the dominant mode. In order to develop capitalism, the integration of the Ottoman social formation with international capitalist relations was absolutely crucial. Hence, the new bourgeoisie played a highly significant role not only economically, but also politically and ideologically to that end.

The Empire was fragmenting and the Ottomans themselves were helpless to prevent disintegration because they could not interpret correctly what was happening. The territories in the Balkans were riddled with sectionalist movements; there was growing disrespect to the Ottomans at the international level which they could not understand, and their internal conflicts were aggravated by the policies of the Tanzimat. In an attempt to control the situation, the Sultan consented to the establishment of the first parliament in 1876. The strategy of the Ottomans in conceding this move was the hope of preventing the further direct intervention of the European powers in the affairs of the Ottoman state. The Sultan was under considerable pressure from Europe to make concessions to the minorities. By setting up a parliament in 1876 it was thought that the rights of both Moslems and non-Moslems could be extended, so that concessions to the people would preserve the powers of the Sultanate. This strategy, however, did not succeed, and the following year Parliament was abruptly dismissed on a pretext. It had become obvious that by setting up institutions which had no roots in the fabric of social relations, the transformation of the Ottoman social formation could not be reserved.

The debacle of the 1876 parliament was not the first nor the last attempt by the Ottoman State to resist the establishment of capitalism. During the period of Tanzimat and especially, in the aftermath of that period, Islam and Pan-Islam became forces to be reckoned with. The Ottomans used Islam to try to restore their position in their colonies and the Sultan used his longstanding right to appoint religious officials in the former Ottoman territories which had come under foreign rule, in order to maintain his influence among their Moslem populations. The Ottoman government intervened and protested vigorously whenever there was any case of maltreatment of the Moslem populations by their new rulers. The Europeans were warned in no uncertain terms that aggression against the Ottoman Empire might lead to a united Moslem uprising against them. Islam thus became an ideological weapon to resist the successive imperialists incursions into the Ottoman realm.

Similar factors stimulated the rise of nationalism within the Empire (Kushner, 1977; Landau, 1981). For the first time, the Ottomans realised that religion alone could not create a modern nation. The Empire had been in decline since the sixteenth century, and an amalgamation of peoples under Moslem control did not provide a firm basis for survival in the face of capitalism. With the increasing penetration of imperialist relationships, the transformation of the Ottoman Empire became more rapid. Although Ottoman ideology decreed that all subjects, regardless of race or creed were equal before the law and loyal to the same dynastic rule, nevertheless non-Turkish Moslems began to agitate for independent nationhoods. While the Ottoman political leadership was effective, and could deliver the goods, Islam served to bind the Empire together, and aspirations to independence had to remain suppressed, but as the decline of the Empire accelerated, struggles for national status came to the surface, and nineteenth century nationalist ideologies could no longer be ignored. As nationalism spread, nationalist ideas which had fired the hearts of Greeks, Serbs, and Bulgarians found fertile ground among Turkish intellectuals, and were freely aired, particularly among professionals and the military. The adoption of nationalist ideas by Turks developed great potential for political action against the Ottoman regime which was now closely identified with foreign interests, and the non-Turkish bourgeoisie. There were many protest groups of different ideological and political persuasions, but they came to share a single label as the Young Turks movement. [7] The movement spread throughout Europe and became very active in opposing the regime of Abdulhamit II. Eventually one of the constituent groups of the movement, The Committee of Union and Progress, was able to force the Sultan first to restore the Parliament in 1908, and subsequently to abdicate.

The Young Turks were products of the westernisation policies implemented during the Tanzimat period. They had been educated in European languages at prestigious schools, and had come into contact with European culture and values. They were aware of developments which were taking place outside the Empire, and also that

> "...Westernisation meant not merely aping the West by adopting superficial reforms, but adopting its economic structure, namely capitalism. They concluded that capitalism required a capitalist class."
> (Ahmad, 1981, p.11)

The Young Turks understood that the social relations of production were undergoing rapid change and that commercial capital was becoming the dominant fraction of the ruling classes (Ahmad, 1980; Toprak, 1982). Since commercial capital was not Turkish, it had to be controlled and to be made Turkish. The achievement of this goal was seen to depend on gaining control first of the bureaucracy. By taking over effective control of state apparatuses, the Young Turks thought that they could undermine the unacceptable face of capitalism, i.e. foreign domination, and instead impose Turkish domination of the general processes of capitalist development. The problem lay not in whether or not to encourage capitalist development, but in how to achieve ownership and control of the means of production, and how to unify the control of economic and political power. They thought that gaining control of political power would lead to the subsequent seizure of economic power, and thus free the Turkish nation from the grip of foreign domination, and remove ownership and control of the means of production from the ethnic minorities.

In 1914, however, the Great War broke out. In addition to the many issues which were already dividing the Young Turks, the crucial question as to whether or not the Ottoman Empire should enter the war became a source of great dispute. I do not think that the Ottomans had any choice but to take part in this battle between the imperialist powers. The Ottoman Empire occupied a highly strategic position and its territories included areas which were bound to be fought over. In addition the war provided a good opportunity to attack Britain and France, who had benefitted most from the humiliating Capitulations, and the policies of the Tanzimat period, and who had established either directly, or through their agents among the ethnic minorities, domination over and the exploitation of the Empire. The Germans skillfully took advantage of this situation and the Ottomans entered the war as allies of the Germans with the explicit intention of breaking the British and French monopolies over the Empire's economic activities.

The outcome of the war, however, was a disaster for the Ottomans. Their defeat resulted in the partition of the Empire, physically in terms of a division of territories, and also in terms of political and economic spheres of activity. Istanbul was occupied, and became the centre for a united allied occupation, with the Sultan retained as a figurehead. By the end of the war, the Young Turks movement was in disarray; there had been too much infighting, too many old scores settled, and too many people had disappeared. The Committee for Union and Progress had ceased to function in 1914. The legacy of the Young Turks, however, the ideas that Turks should unite in order to survive, and that Turks should benefit from the established process of capitalist development, remained strong. These nationalist sentiments became the basis not of a new intellectual movement, but of a desperate national struggle for survival in the face of invasion by European forces. In the wake of military defeat in 1918, Britain, France, Greece and Italy launched invasions with the aim of dividing the remnants of the Ottoman Empire between them, and Anatolia took the brunt to these attacks. Led by Mustafa Kemal, the War of Independence was launched in 1919 to liberate the country from the foreign invaders. Two congresses held in Erzurum and in Sivas in 1919 established the main class alliances on which Kemal's forces depended, and the Grand National Assembly was established in Ankara in 1920 as the political platform to unify different interest groups taking part in the struggle for independence under one banner. [8] The main support for the independence movement came from the nascent Turkish bourgeoisie. This national liberation struggle ended in victory for the Turks, and the new Republic of Turkey was established following the signing of the Lausanne Peace Treaty in 1923, a treaty which also abolished the greatly resented Capitulations (Meray, 1970, 1972, 1973).

Mustafa Kemal became President of the new republic and also its main ideologue. He claimed unequivocally that the Turkish Republic was Turkish and that the Ottoman Empire had been Ottoman, so that Turks were not responsible for the deeds of the Ottomans. [9] This world view constituted the ideological basis for the economic, political and social policies which were to be implemented later. On the basis of the same ideology, the Turkish social formation became an inward-looking totality, deliberately closed to the outside world. At the same time it was seen to be essentially for Turkey's development that economic and political links with the West were maintained (Kazancigil, 1981). This new ideology of the Republic of Turkey marked a deliberate break with Ottoman history, and the Empire's experiences of humiliation and exploitation. Turkey was to be a Western, secular republic, but a distinctively Turkish one in terms of culture, and an independent sovereign nation-state. Mustafa Kemal swiftly and efficiently imposed important symbolic changes on the Turkish people, such as the adoption of surnames, and hats in place of the "fez", but he also introduced Roman script in place of the Arabic script used by the Ottomans, and also the Gregorian calendar instead of the Moslem calendar. These changes were seen as essential for facilitating communications between Turkey and the West and were a decisive factor in differentiating Turkey from the rest of the Moslem World and especially the Arab World. Mustafa Kemal was given the unique surname Ataturk (the father of the Turks) by the Grand National Assembly in order to emphasise the appreciation and the gratitude of the Turkish nation. This gesture by the national assembly of the new Turkish Republic also meant that Ataturk, as he came to be known, was given totally free hand in determining the future of the Turkish nation. In the history of the Ottoman Empire the Sultans themselves had very rarely enjoyed the amount of power that Ataturk had during his reign as the founder and the leader of the Turkish Republic.

Mustafa Kemal Ataturk was faced not only with constructing a new Turkish state, but also with generating new political, economic and ideological practices which would fulfil the conditions of existence for the accumulation of Turkish capital. In the next section I look at etatism as the key strategy of the Kemalist regime for capitalist development in Turkey.

ETATISM AS THE STRATEGY FOR CAPITALIST DEVELOPMENT

Etatism was one of the six principles (Karal, 1981) which Mustafa Kemal laid down as being necessary for the birth of a new Turkish nation from the ashes of the Ottoman Empire. The other five principles were Republicanism, Nationalism, Populism, Revolutionism/Reformism, and Secularism. These six principles became known as Kemalism, and became the basis for the party programme of the newly formed Republican People's Party (RPP). The RPP was formed and led by Mustafa Kemal, and this party was thought to be adequate for the representation of national interests, so no other parties were formed. The emblem of the RPP was six arrows representing the six principles of Kemalism. The six principles were also written into the 1937 constitution, and became the guidelines, not only for the Kemalist regime, but also for successive governments up to and including the present military regime. The looseness of their definition has allowed them to be used to legitimate any action taken by the capitalist state. Although all six have been important for the course taken by the development of capitalism in Turkey, I shall only deal with etatism, as being the most crucial influence on the transformation of the Turkish social formation.

The war had taken its toll. Although the Turks had won their independence, the war years had shattered the economy, and devastated the Ottoman industrial base. Anatolia, where the new republic was to be located had been ravaged. The Turks had to rebuild the economy almost from scratch, but this time there was not going to be any separation of economic from political power; the Turks were going to have both. The lessons of the Ottoman experience had been well learned. The new leaders were heavily influenced by the ideas of the Young Turks movement, and they finally had the opportunity to build a nation on the basis of these ideas which had been absorbed into _Kemalism_. The ethnic minorities had either fled or been "pacified". Some members of the ethnic minorities, however, both remained and survived, particularly in Istanbul. Those who were willing to accept the new political regime were allowed to operate as Turks, and continued to dominate commercial capital for a considerable time. The new regime was intended to be a clear manifestation of the reaction of a nation to a very long period of imperialist exploitation by European capital and their agents. Thus the determination of the political, ideological and economic practices of the _Kemalist_ regime was geared to the development of an independent, modern and capitalist Turkey.

Ataturk saw that the successful development of capitalist Turkey depended on certain programmes being implemented by the general agreement of the people as a whole (Ahmad, 1981). The Republican People's Party was set up precisely for the purpose of providing the people of Turkey with a platform for participation and representation at a national level (Tuncay, 1981; Kili, 1976; Bila, 1979). But a key question was that of which people were to have access to the distribution of political power; it was certainly not going to be the peasants of Anatolia! The social classes which were to be represented at the Grand National Assembly were the landlords, the merchants and professionals. The aim of the assembly was to provide a means of bringing together those social classes which had supported the war of independence, on the clear understanding that these classes would benefit most from the new order, when it was created. This was the time to establish a co-ordinated scheme to ensure the smooth completion of the processes of capitalist development. The state, the sole proprietor of political power, could oversee the general development process and wherever and whenever necessary, could intervene to correct the operation of market mechanisms, and to generate new impetus for the accumulation of capital. At the same time, the development of private capital was urgently needed. The role of the state was to safeguard and create the conditions of existence for the reproduction of capitalist relationships, and to inhibit the reproduction of non-capitalist relations of production,so that the development of private capital could be promoted, and the Turkish social formation effectively transformed. The state, therefore, had certain tasks to perform which were independent of private capital, and yet in accordance with the interests of private capital.

In the following sections, I look at the development of class structure and of the contradictions within the power block as various applications of _etatism_ were developed. _Etatism_ meant that, (i) the state had to perform certain economic functions, such as infrastructural investment, which were essential for the development of private capital, and also (ii) that the state had to intervene in those spheres of economic activity where private capital was not in a position to play the active role, either because of lack of investment funds, or because of the impossibility of a quick return on investments. The necessity of these economic practices meant that the state had to create corresponding juridico-political, and ideological structures. It was this correspondence between economic, political and ideological levels which created and reproduced suitable conditions for capital accumulation. These _etat-_

ist policies laid the foundations for the later adoption of the import-substitution model of capitalist development. [10] I argue that the choice of the import-substitution model was the logical outcome of the conjunctures which proceeded it, in which etatism dominated the form taken by emergent Turkish capitalism.

THE CONSOLIDATION OF THE REGIME: 1923-1929

During this period, the identification of political and economic power became well and truly entrenched. Etatist policies as envisaged by Ataturk and his associates, were put into practice with the assistance of indigenous commercial capital. The political leadership consisted mainly of people with military and bureaucratic backgrounds, but many of those who held key positions of authority and power were actively engaged in economic activity, both public and private, alongside the commercial bourgeoisie.

In periods of the partial transformation of a social formation into capitalism, the class structure may not be well developed, and hence the nature of class practices may not be particularly clear. When the ideological and legal frameworks are correspondingly ambiguous, the resolution of contradictions may turn out to be a relatively easy task for the state. The initial period of the implementation of etatist policies in Turkey is a case in point. The etatist programme was fairly quickly put into effect because classes were not as yet clearly developed. The state encouraged the development of commercial and industrial capital, but non-capitalist relations of production were, nevertheless, being represented at the political level. The potential domestic market was so vast, and the potential opportunities for development were so abundant, that the expansion of capital could take place, up to a point, without endangering non-capitalist production relations. Turkey needed everything and anything that could be produced.

Because of the urgent need to build and to develop a viable economy with a corresponding political structure within the framework of Kemalism, it was felt that the problems of developing Turkey had to be discussed in a forum which would produce concrete proposals. Thus the Ministry for Economic Affairs organised the Izmir Economic Congress in 1923 (Okcun, 1970). The composition of this congress is particularly interesting, because it shows the class formation of the new Turkish Republic. Delegates to the congress were explicitly elected on the basis of their professions or their class origins. Workers and peasants attended the congress alongside the representatives of commercial capital and landed interests, but the outcome of the congress entirely favoured the big merchants and big landlords. The group which played the most vocal role in the congress was the Istanbul merchants. They wanted to take over the businesses left behind by Greeks and Armenians, and they also wanted to expand into other spheres of economic activity. In fact, Avcioglu suggests that the Istanbul merchants had wanted to organise a similar congress in Istanbul, but the government resisted pressures from Istanbul for fear, one presumes, that the rest of the country would be suspicious of Istanbul, already the most developed area, being favoured at their expense. [11]

It was evident, however, that Istanbul's commercial bourgeoisie was by far the most developed and articulate fraction of the ruling classes, and they knew exactly what they wanted from the state. In 1922, they had formed the National Turkish Trade Association (NTTA). NTTA's manifesto stipulated certain specific aims. These were:

> a) to ensure the superiority of Turkish merchants in either import and export or wholesale and semi-wholesale trades.
>
> b) to work for the creation of consortium, trusts and similar trade organisations among the Turkish merchants according to their specialisation.
>
> c) to work for the acquisition of state protection and control over those organisations working in the import and export trade.
>
> d) to follow the events which would have an impact on the nation's trade and thus to provide guidelines for the government to act according to the interests of the merchants and the nation at large.
>
> e) to inform the outside world of the Turkish world of trade and about Turkish firms, and to inform Turkish merchants swiftly of economic, financial and commercial events which are taking place in the West. (A.H. Basar quoted in Avcioglu, 1968, p.166)

The other aims stipulated in the manifesto were that foreign capital which was to exploit Turkey's mineral resources should receive guidance from the state which should also participate in these enterprises and ensure that Turks became the majority shareholders, and generally consolidate co-operation between Turkish merchants, and represent their interests.

The sections of this manifesto which dealt with the role of foreign capital in Turkey's economic development were adopted by the Izmir Economic Congress. Although there were a fair number of workers and peasants at the congress, there was no real acceptance of trade unions or the right to strike. [12] The message that came out of the congress was that the "national economy" had to be built and developed. Its leadership had to come initially from the state, but without jeopardising the interests of private capital. In the meantime, the ruling classes were going to co-operate with each other so that the power block could be maintained.

The role of the political leadership turned out to be twofold. Firstly, in the person of Ataturk, the regime had its founder and undisputed leader. Ataturk was very much in tune with the expectations of the Turkish bourgeoisie and so could interpret the demands of the ruling classes to the workers and to the peasants of Anatolia in a very articulate and acceptable fashion. In his opening address to the Izmir Economic Congress, he said that he was addressing and audience made up of farmers, artisans, merchants and workers. He asked rhetoricly which one of these groups could possibly be superior to the rest, and he went on to say that no one could deny that each of these groups needed each other. [13] This theme became part of the dominant ideological framework of the congress and there was no overt recognition of differences of interest either within the ruling classes, or between the ruling and subordinate classes, for some considerable time. Turkey was to build an economic system unique to its specific conditions and to meet its own needs. It was not going to be corporatist, as advocated by Mussolini, nor would it become socialist. It was going to be a mixed system whereby its economic enterprises would be partly state owned and partly privately owned.

The second role of the political leadership was then to decide which sectors of the economy were to be incorporated into the framework of state economic activities, and which were to be left to private enterprise. Here the political leadership had a very delicate situation to deal with. Their criterion was that the state should develop those sectors of the economy which required the establishment of major industrial enterprise which the private sector could not finance. Elsewhere private capital should be encouraged to develop and expand by being left to its own devices (Bulutoglu, 1970; Goymen and Tuzun, 1976; Uras, 1979). The state should provide the necessary politico-legal framework without which private capital could not flourish, while the state economic enterprises should ensure balanced industrialisation and economic growth.

These policies, however, sowed the seeds of contradictions between industrial and commercial capital which developed in the years to come. These developing contradictions were not apparent at the time, and were not recognised by the Istanbul fraction of commercial capital. Even when Ataturk himself privately established the Business Bank (Is Bankasi), this did not alert commercial capital to the dangers which threatened them (Silier, 1975). The Business Bank was set up in 1924 specifically to provide financial support for the Turkish bourgeoisie to develop factories and businesses. [14] The establishment of the bank was the end product of a process instigated by the Turkish state to buy up remaining foreign businesses, and to generate a strong state-backed and Turkish-owned industry. It was, therefore, publicly controlled, but privately owned. The Young Turks had encouraged industrialisation by tax concessions and customs exemptions for imported machinery, but what they had developed was largely destroyed during the war, so the task of reconstruction facing the leaders of the new Turkish state was enormous. The Business Bank started by investing in small businesses, but very soon made heavy commitments to the development of coal mining on the Black Sea coast, as coal was chosen to be the major energy source for the industrialisation process. The state also established the Turkish Industrial and Mining Bank to provide financial backing for state enterprises. Industry was further encouraged by the legalisation of the Chambers of Trade and Commerce set up under Sultan Abdulhamit II. These were originally given the tasks of developing crafts and skills, providing funds for training, settling disputes among workers and generally overseeing the running of industrial and commercial enterprises. They were now given an extended economic role and new importance.

The most important piece of legislation which passed through the Grand National Assembly was the "Law for the Encouragement of Industry", in 1927. According to this Law:

> a) factories and mines were to be given free land to build on.
>
> b) these enterprises and the land they developed would be exempted from taxes.
>
> c) if the necessary materials for building or expanding these enterprises could not be obtained in Turk-

ey, then imports of such materials would be duty-free.

d) the goods, machinery and tools for these enterprises would be carried by rail or sea at a 30 per cent discount.

e) the Council of Ministers could decide to give bonuses to these enterprises equivalent to the value of 10 per cent of the value of the commodities they produced.

f) if required, the enterprises could have a reduction in the price of salt, explosives and white spirit.

g) state institutions, local authorities, municipalities and their agents had to buy and use Turkish products even if foreign products were up to 10 per cent cheaper.
(Avcioglu, 1968, pp.184-85)

Employment had to be limited to Turks, with foreigners only employed in exceptional circumstances. Those industrialists with factories or other industrial enterprises who benefitted from this law were granted monopoly status in their sector of commodity production for 25 years, free of state intervention.

With the introduction of this "Law for the Encouragement of Industry", the industrial bourgeoisie was given economic priority over the whole of the commercial bourgeoisie, and was certainly in a much better position to expand than were the landed interests. Agriculture, however, was not ignored during this period of consolidation, and it remained the largest sector in the Turkish economy. The incorporation of the Turkish economy into international trade relations in the late nineteenth century had long lasting effects on the development of agriculture. The Lausanne Treaty of 1923 had imposed a liberal foreign trade regime on the new Turkish Republic. The prevailing terms of trade for agricultural sector provided the impetus for growth in the national economy and as a result, agricultural and land-based interests became important and indispensable factors in the determination of the political and economic restructuring of the social formation. These interests were further favoured by the reorganisation of the Agricultural Bank (Ziraat Bankasi), in 1924, to meet the increasing demands for credit coming from the big landed interests, and by the import of tractors which improved the efficiency of land use and increased agricultural productivity substantially.

Another important decision taken by the state was to abolish the land tithe in 1925. Instead, a new tax on produce set at only 6 kurus in the thousand was levied, and eventually this tax, too, was discarded and replaced by a tax levied on agricultural income. [15] As a result, the agricultural sector's contribution to state revenues decreased sharply, which led to a corresponding increase in the tax burden on trade and industry. This taxation policy made the development of industry and the growth of urban areas much more difficult, as well as being in contradiction to industrial development policies.

> "Doubtless an important element in the decision to abolish it [the tithe] had been the implicit alliance which had been formed between the government and the large farmers during the War of Liberation...There was...also an element of concession given to the large farmers to secure their complicity in the face of superstructural reforms by the bureaucracy during this period."
> (Birtek and Keyder, 1975, p.451)

This was a decision which the new regime had to take in order to be assured of the support of the large and powerful landed interests. Given the nature of Kemalist policies, the transformation of Turkish capitalism required a fundamental breakdown of the old, non-capitalist modes of production. Although mechanisation of agriculture was bringing about much more efficient use of productive forces, the non-capitalist relations of production were still being maintained, and the capitalist transformation of agriculture was still to come. It was only later that the growth of industry and the subsumption of direct producers to capital changed the nature of commodity production in agriculture, and transferred political and economic power from the countryside to the cities where the greater percentage of the gross domestic product was then produced.

Although the chief aim of etatism had been to develop industry, in practice other considerations had to be given more immediate priority. The fierce contradictions which were to bedevil Turkey at later stages of capitalist development had not yet fully emerged, but the pre-conditions for these confrontations were laid down during the period 1923-29. I would argue that Ataturk and his government knew exactly what they were doing, but the need to obtain the support and co-operation of the landed interests and, to a certain extent, of the small and middle peasantry, outweighed other considerations. This situation was a direct outcome of the Izmir Economic Congress, since the structure of the power block gave rise to the pursuit of contradictory policies in industry and trade and agriculture. Politics dominated economics in the formative years of 1923-29. The issue of whether industrialisation should take priority over the development of agriculture and trade, led to battle lines being drawn between industrial interests and commercial and landed interests. In the next section, I argue that the ensuing struggles had important repercussions for the transformation processes but that, since the fractions of the bourgeoisie were still not clearly developed to the point where one or more fractions could establish a monopoly over the use of state power, the state was able to play a relatively autonomous role, and to prevent outright conflict.

CRACKS IN THE POWER BLOCK: 1929-39

The period 1923-29 was one of consolidation for the new regime, but their clear objectives were undermined by confusion and contradiction in their policies. This was a very effective phase of capitalist transformation at the level of political practices, but the impositions of the Lausanne Treaty added to contradictions at the level of economic practices, by encouraging an influx of foreign capital and thus undermining the Kemalist goal of Turkish economic development based on industrialisation. This situation came about because Turkish private capital had little incentive to invest in the development of Turkish industry when quicker and easier profits could be made under the liberal foreign trade regime imposed by the Lausanne Treaty, through import and export trade.

Given the degree of maturity displayed by the commercial bourgeoisie in directing state activities in its own interests, the dilemma facing the industrial bourgeoisie and the political leadership worsened. What had in effect taken place during the previous period of the Republic was that the commercial capital based in Istanbul had replaced the minorities, and especially Greek and Armenian capital. Foreign investment by the European powers continued, and the Anglo-French owned Ottoman Bank monitored and handled the flow of foreign capital into the country. Its Turkish equivalent, the Business Bank, founded only in 1924 with relatively meager funds, was trying to stimulate the growth of investment in Turkish industry with little success. [16] The import business was turning out to be so lucrative that Turkish capital could not be diverted into medium term industrial investments. The difficulty of creating strong industrial base and hence strong industrial bourgeoisie, were increasing. The commercial bourgeoisie was reaping the benefits of state policies in trade, coupled with non-restrictive trade agreements imposed by the Lausanne Treaty. Agricultural capital was also at a relative advantage, at least temporarily, in being able to expand its productive capacity for domestic as well as export markets by turning to mechanised farming.

The only sector which seemed to be losing out, despite the explicit declaration of national economic priorities, was industry. Although there had been some investment in industry by private capital, this was not nearly enough for the effective expansion of the industrial sector under the prevailing economic and political conditions. In 1929, the consequences of a number of problems in the emergence of Turkish capitalism came to a head. Imports were exceeding exports quite substantially, and the trade deficit was aggravated by the decline in the price of agricultural commodities, while foreign exchange for the purchase of capital goods dwindled rapidly. The Ankara government found itself at a crossroads. As it had committed itself unconditionally to the generation of new impetus for the economy wherever investment was lacking, its only alternative was to increase state control of industrial production and of the economy as a whole. This decision to increase state intervention in the economy was further influenced by the world recession of 1929. [17] Hence the path taken by Turkish capitalism at this point was dictated by the specificity of the Turkish social formation, but this also embodied the contradictions of world capitalism. Once again imperialism, as shown by the international division of labour, determined the transformation processes of Turkish capitalism as one of its constituent social formations. The outcome of this conjuncture for Turkish capitalism was the taking of a deliberate decision to create and to protect domestic industry by closing the Turkish economy to world markets. The chosen path of Turkish capitalist development was to be based on an import-substitution model of production, coupled with a strong corporatist state. [18]

The 1923-29 period had given the Turkish state the opportunity to consolidate its dominance by developing and diversifying its apparatuses for the penetration of major levels of social life. The economic, political and ideological practices of social classes had been incorporated, to a very large extent, into the operational framework of the state. The structures of domination and subordination had been established. The bureaucracy had become very influential in centralising the distribution and reallocation of resources, not only at the level of economic practices, but also affecting political and ideological practices. The weak development of the class structure, however, meant that no fraction or fractions could yet produce dominant class practices to combine economic and political power and thus to dominate the development of capitalism. This situation is not surprising when one considers the nature of capital accumulation up to this period. The state had, for every practical reasons, lent its full support to the development of an industrial bourgeoisie, in the face of very articulate opposition from commercial and agricultural capital. Now that the state was well established and state power could no longer be threatened internally, the state could afford to ignore this opposition and to pursue policies which favoured the industrial bourgeoisie at the expense of the other fractions of capital.

The important decision to close the economy and to stimulate industry was forcefully articulated by the political leadership in terms of the protection of the "national economy". In order to achieve their aims, certain co-ordinated changes in the political, legal and economic structures had to be made, changes which had long lasting effects on the development of Turkish capitalism. These changes can be classified into two groups.

1) The first task of the state was to change the liberal foreign trade regime and fiscal and monetary structures, in order to protect Turkish interests. In 1929, Ismet Inonu, the Prime Minister, said that wheat production, Turkey's major export crop, had to be protected. He also urged the protection of industrial interests, and especially textile industry. [19] In order to achieve some protection for industry, customs duties were introduced. Later on, in 1930, the Central Bank of the Turkish Republic was founded with the explicit aim of withdrawing the right to control and issue currency from the foreign-owned Ottoman Bank. This decision gave the state the power to control and regulate its own monetary and fiscal policies to ensure the development of a strong financial sector and to safeguard the stability and value of Turkish currency. In the same year and in the following year, further legislation was passed to close the economy further to world markets and, more specifically to foreign investments. [20]

The first legislation to pass was the "Law for the Protection of the Value of Turkish Currency". According to this law, anybody or any organisation who, in the pursuit of private gain should devalue Turkish currency, would be severely punished. The Council of Ministers was empowered to take measures to protect the stability of the currency, but the nature of these measures was not specified. The only comment on this ambiguous law at the time came from the Prime Minister Inonu, when he said that the government was going to exercise its rights with great discretion, and the bill passed through the Grand National Assembly without a murmur of opposition. The second measure to be passed, also in 1930, was the "Law to Prevent the Adulteration of Trade and the Control and Protection of Exports". This law was intended to improve the quality control of exported Turkish made goods. Since there were no legal restrictions on the export of commodities, adulteration and trade malpractices were common. Such practices were to stopped in an effort to increase international confidence in Turkish products. There was no opposition to this bill, either. The third measure, passed in 1931, was concerned with the restriction of imports, and its title is more or less self-explanatory. This was the "Law Prohibiting, Restricting and Applying Conditions to Imports coming to Turkey from States whose Trade Transactions with Turkey do not result in a *modus vivendi*".

Boratav's interpretation of the implications of these three laws reflects the contradictions developing at the time. [21] He sees the first two laws as reactions to the role played by dominant capitalist interests. The first was passed to control speculators, and the second to control importers. The third was passed to restrict further the fundamental interests of the importers, by introducing protective customs duties in order to end the non-restrictive trade practices enjoyed by this fraction of capital as a result of the Lausanne Treaty. I do not agree, however, with the argument that the passing of these laws marked the start of the *etatist* period. I would argue that the foundations of *etatism* were laid down between 1923 and 1929, and the introduction of legislation regulating Turkish foreign trade only made existing *etatist* policies much more widely applicable. As a result of these policies, the contradictions between the industrial and the commercial bourgeoisie surfaced, and the relative autonomy of state became a vital determining factor in the resolution of class conflict, and hence in the maintenance of the power block.

2) The second change came in the form of state economic policies. Between 1929 and 1939 two five-year plans for economic development were implemented. [22] The First Five-Year Plan in 1934 saw the development of industries for the production of chemical products, iron, paper, sulphur, sponge, cotton and wool textiles. The long established sugar industry was also encouraged, and new factories and mills were located near their sources of raw materials. In 1933, *Sumer Bank* was established to extend credit to light and heavy industry, and to create new state economic enterprises, which were 51 per cent state owned and 49 per cent privately owned. The production of cotton, wool, leather goods, carpets, coke and cement all came under state economic enterprises, and these enterprises were virtually monopolies in their own markets. *Sumer Bank* was not a bank in the usual sense of the term, but was a successful state enterprise.

Ataturk also set up the Hittite Bank (*Eti Bank*) in 1935 to co-ordinate and to develop domestic natural resources for productive use. In the same period several other enterprises were set up by the state. The Real Estate Credit Bank (*Emlak ve Kredi Bankasi*) was formed to operate as a financial base for giving credit to both public and private construction projects. In 1933, the Provinces Bank (*Iller Bankasi*) was founded to encourage provincial projects at village and municipal levels.

By this time, the outlook for the future of Turkish capitalism had become fairly clear. Following the great depression, a large number of Turkish firms had gone bankrupt, and unemployment was reaching serious proportions. Long established trade links were being severed, as a result of the international recession, as well as because of changes in foreign trade regimes. Available capital resources were being reallocated by the state intervening directly in the economic sphere, with the aim of developing Turkey's industrial potential. By forming new industrial enterprises and placing them in the hands of private capital, the state was creating the conditions of existence for the development of an industrial bourgeoisie. In doing this, the state was simply implementing *etatism* as one of the six principles of *Kemalism* formulated at the time of the foundation of the Republic. The 1927 "Law for the Encouragement of Industry" had specified very clearly the bases for state intervention in the economy, but when the state actually began to put this law into effect, those whose interests were threatened, started to protest. The commercial bourgeoisie in particular was very vocal in its criticisms of the new political and economic measures. [23] They saw themselves as the backbone of the Turkish economy, and liberalism as their ideological guide. The conflict between liberalism and etatism, however, was not really very clear cut. On the whole, as the aims of the National Turkish Trade Association show, the commercial bourgeoisie did accept state intervention as necessary for capitalist development. The real opposition from the commercial bourgeoisie was not to etatism as such but to the high port taxes and high tariff barriers which were making the maintenance of free trade difficult.

The development of this opposition took a political form with the formation of the Free Party in 1930, ostensibly to give the people choice of representation, but essentially to represent commercial interests, especially those of the importers, and to publicize and challenge the etatist policies of the RPP. [24] The growing industrial bourgeoisie, on the other hand, allied itself with the RPP. In the case of the alliance of industrial and agricultural capital, and the state, commercial capital retreated fairly quickly, and the multi-party experience was a short-lived one. [25] One important consequence of the formation of the Free Party though, was that the state was alerted to the problems of organising capitalist development in the midst of severe contradictions and growing class struggle. Thus the formulation of subsequent policies was marked by increasingly aggressive state practices (Tuncay, 1981).

Further progress was made in the implementation of etatism during the latter half of the 1930s, especially following the Second Five-Year Plan introduced in 1938. The Land Products Office was established to maintain the stability of agricultural prices by buying and selling according to variations in production and in season. The State Monopolies Company was set up to manufacture and control tobacco products, alcoholic drinks, spirits, matches, tea and salt. The Business Bank, founded in 1924, became heavily involved in the development of railways, lumber, coal, sugar, textiles, electricity and insurance. Extractive and heavy industries were also established during this period, in the form of coal mines, steel mills, factories for agricultural machinery and meat processing, electrical plants, chemical plants, aluminium smelters etc. Simultaneously, efforts were being made to raise the level of agricultural exports.

The state, therefore, in implementing etatist policies, was not trying to discourage the development of private capital. The advocates of etatism put forward two main arguments. Firstly, they argued that high protective trade barriers and the introduction of exchange controls provided an environment for the expansion of Turkish economic activity, and the most important economic activity was industrialisation. Protective trade barriers kept foreign investments out of the national economy and prevented Turkish capital from having to compete with more advanced capitalist economies. The second argument, which followed from the first, was that private capital could not fulfill all the requirements of industrial development. It certainly did not have the capacity to develop those areas of commodity production which required long term investments, possibly yielding low returns.

These two arguments determined the nature of etatist policies during this period. Railroads, electricity generating plants, iron and steel mills, roads and maritime fleets were infrastructural developments which private capital needed but which it could not establish itself. In laying down the foundations fo the growth of the industrial sector, the state was determining the course of the subsequent phases of capital accumulation. Agricultural prices were affected by increasingly adverse terms of trade vis-a-vis prices for manufactured goods, export markets were disappearing fast and international credit opportunities were drying up. The encouragement of industrial production, however, was being made possible by these infrastructural developments and the production of raw materials. Since private capital could not or, in many instances, would not invest in these areas, the state could only achieve its declared aims through _etatist_ policies.

The role that the state was compelled to play was subject to various interpretations, and thus to different forms of legitimation. The famous _Kadro_ movement, which was gathered around the publication _Kadro_ (Cadre) from 1932-34, for example, defended the role of the state in fighting "capitalism" (Aydemir, 1968). _Kadro_ saw capitalism as being identical with the interests of private capital, and thus argued that private capital had not contributed to Turkey's development, so that the state was justified in taking over key areas of production. Despite these simplistic and misleading view, the _Kadro_ movement played an important part in the legitimation of the Kemalist regime. [26] This was, however, a much more cautious justification of _etatism_ than that of _Kadro_, with the implication that as private capital developed, the role of the state could diminish, a position that was to become significant later with the rise of the Democrat Party. Bayar argued that _etatism_ was based on the principles of private ownership of property and work for personal gain but, nevertheless, _etatism_ was committed to the idea that where private capital failed, the state would give its support. [27] This general view of the vital role of private capital in the transformation of Turkish capitalism was publicised in order to calm the fears of the ruling classes and to indicate the state's view of its own role in the development process.

The years leading to the Second World War were the years of what can be loosely termed as mending the fences within the power block. The outbreak of the war brought changes which upset the fragile balance of power to such an extent that the previously all powerful RPP found itself in dire need of popular support. Ataturk had died in 1938 bequeathing a legacy of a corporatist state, based to a very large extent on the use of personal power and nepotism. His Republican People's Party was, however, not a party of the people. It was a party of the bureaucratic and professional strata which not only lacked any popular base but which actively feared the people. [28] By controlling the bureaucracy and identifying itself very closely with the state apparatuses, but once Ataturk was no longer there to exercise his tremendous personal authority over the masses, this situation could not continue. There was, however, no immediate response from the masses. Opposition to the RPP came from within its own ranks.

The only effective opposition to the regime that could have come from outside the RPP would have been from the rural areas, but the rural classes were quiet and acquiescent. The landed interests were part of the power block dominating the transformation processes, while the poor and middle peasants had no ideology or organisation adequate to challenge the _Kemalist_ regime. There had been some resistance to the social reforms introduced by Ataturk, but the widely feared rural gendarmerie were well able to maintain the authority of the regime. The leadership of the RPP was well aware that the rural classes posed a potential threat to the existing order. This threat did not come from unorganised peasants, but from the possibility of landlords being able to mobilise the peasantry through feudal obligations. To prevent this threat being realised it was thought essential to alter the balance of power between the urban and rural classes, which meant that the dominance of agriculture in the Turkish social formation had to be undermined. The state, dominated by the RPP, had taken the decision to close the economy. [29] This decision had the effect of putting agriculture at a disadvantage; it affected the import of tractors and the export of agricultural products, thus confining agricultural development to the domestic market. In the long run the regime hoped to incorporate agricultural production into the processes of generalised commodity production, through the development of exchange and distribution relationships. [30] The result of this was to be that the peasants would become wage-labourers, and that those who could not be absorbed into agricultural wage labour should provide a source of labour power for the new manufacturing sector. To this end, the legal framework for the creation of the labour market was prepared with the enactment of two laws in 1932 and 1936. [31] These intentions, which favoured industry at the expense of agriculture, had the effect of turning substantial sections of the landed interests away from the RPP, and had important consequences for the development of class struggles during and after the Second World War.

Turkey remained neutral during the war, although the development of Turkish capitalism was inevitably influenced by changes in world conditions. In the following section I look at the impact of the events of the war years on the development of Turkish capitalism, and argue that it was during this period that an alliance between commercial and agricultural capital was formed. The outcome of this alliance was that for the first time, state apparatuses were penetrated by particular class fractions which were able to monopolise state power.

THE WAR YEARS AND THE RISE TO DOMINANCE OF COMMERCIAL AND AGRICULTURAL CAPITAL: 1939-1950

The accelerated implementation of *etatist* policies had not only strengthened the *Kemalist* regime, in spite of the erosion of its popular base, but had also created political and economic structures which were largely free from external determination. The exceptional circumstances of the war years, however, began to effect Turkey's relationship with world capitalism and thus its internal development. Turkey was in a state of semi-mobilisation; a large proportion of the working population had been conscripted into the army, and defence expenditure was increasing. Because of shortages on the world market caused by the war, there was a sharp decrease in the import of capital goods used as fixed capital or intermediate products in the manufacturing sector, and taxes and duties levied on imports declined proportionately. At the same time wheat production declined to an alarmingly low level as did the production of other agricultural produce, leading to a danger of food shortages which would make the maintenance of a large army and large urban populations very difficult. The state needed strategies for maintaining industrial and agricultural productions, but decreasing incomes, and falling levels of production reduced revenue from taxation, thus reducing state resources for financing new public expenditure. It was a classic example of an inflationary situation. The outcome was shortages of daily necessities, rising prices, falling real wages and the inevitable profiteering and blackmarketing. Falling levels of agricultural production coupled with high international demand for food created by the war, meant that there were opportunities for the export of food crops which further reduced the supply of food for the domestic market and led to a few black marketeers effectively controlling distribution. At this point, state intervention took a different turn with the introduction of a new "Law of National Protection" in 1940. [32] This law marked a new stage in the development of _etatism_, by giving state bureaucracies new and sweeping emergency powers for the control of the economy which, _de facto_ could be exercised on a discretionary basis. This led to a very uneven implementation of the law, and the forms of capital accumulation which resulted throughout the war years, and also afterwards, were shaped by the uneven state practices which followed.

The powers given to the government by this law were unprecedented. On the industrial front, the government was to determine the nature and quantity of the output of industrial and mining enterprises; all investment projects would be subject to government supervision, and those enterprises which did not abide by the decisions of the government would be nationalised in return for suitable compensation. An amendment made in 1942, allowed for the unconditional expropriation of industrial and mining enterprises if the government deemed it necessary. All means of transport could be allocated to areas where they were especially needed, and the government could buy vehicles as and when they were required for any purpose.

On the agricultural front, the government had similar powers. It could expropriate land over 500 hectares in return for suitable compensation, if it saw fit, and run it as a state economic enterprise. The government could also purchase agricultural commodities at prices well below the market price and sell them at higher prices. Wheat, as the staple food of the population, was rationed and sold cheaply, as was coal the main source of energy. With the 1942 amendment, agricultural producers were no longer required to sell their products to the government, if a private merchant could offer them a higher price. Once farmers were selling to merchants rather than directly to the government, the bureaucracy had the power to decide which merchants should be favoured in selling to the government or to state economic enterprises, and thus how far wealth could be concentrated and monopolies created. The situation was similar in industry, but the development of the agricultural sector was critical at this period because the people urgently needed food, and the army, always a potential hazard, had to be adequately fed and clothed. The _Kemalist_ regime had its repressive apparatuses to maintain order, but Turks had had too much experience in the running of states for the regime to expect to be able to survive indefinitely on the basis of repression alone.

Although the government now had powers to control prices, to intervene as a buyer and, if necessary, to import goods to satisfy national demands, these powers were used discriminately to determine the process of capital accumulation. By giving huge contracts to commercial capital, the state encouraged the concentration of wealth in selected hands. The alliance between commercial and agricultural capital subsequently became a dominant force in altering the balance of power, supported by bureaucratic elements in pursuit of personal gain. These personal interventions by bureaucrats had, therefore, considerable impact on the way the law implemented, and in particular on the way decisions were taken over where and when the state should intervene, and where and when capital should be encouraged. In 1942, a change of government actually encouraged the development of private capital without bureaucratic interference, although the 1940 "Law of National Protection" was still in force, which encouraged the processes of accumulation that were already in motion.

In 1944 a wealth tax was introduced which was intended to tax the super profits made by commercial capital, and an equivalent tax was also introduced on agricultural produce. The aim of these taxes was to create state revenue which could be used for public expenditure programmes, and also for the import of capital goods for the industrial sector. Taxing the wealth created by commercial and agricultural capital, however, was bound to be ineffective, and the wealth tax was in fact thoroughly abused. The bureaucratic elements who operated the allocation and redistribution of funds, credits and contracts were involved in private transactions themselves, and it was in their interests to help evade the requirements of the wealth tax. The Director of the Financial Administration of Istanbul wrote an account of the implementation of the wealth tax in Istanbul Province, and concluded that there was widespread abuse, as not only had the tax failed to prevent the accumulation of gross wealth, it had actually contributed to its development. [33] During the war, the distribution of income underwent considerable changes, and merchants, industrialists, and those fractions of agricultural capital which could keep up with the rapidly changing situation, benefitted greatly from the uneven implementation of _etatist_ policies. The wealth that was accumulated during this period was later to form the source of investment in industry and agriculture.

This new wealth had been encouraged by the opportunities for the export of agricultural produce and raw materials, which were desperately needed by the embattled imperialist powers. The accumulation of wealth together with the soaring inflation rate, led to the commercial bourgeoisie making demands for the re-introduction of liberal economic measures, and for the opening of the Turkish economy to world markets. The political and ideological climate changed rapidly in sympathy with these new demands which saw future economic development as dependent on an alliance between domestic private capital and foreign capital, rather than as between domestic private capital and the Turkish state. Turkey had agricultural products and natural resources to offer, and in return could offer a market for the manufactured goods of Western capitalism. Opening the Turkish economy would mean welcoming in foreign capital to invest in Turkish development, while the role of the Turkish commercial bourgeoisie would be partly to act as agents for this foreign capital, and partly to provide the necessary trade relations for the export of agricultural products. This view of future development meant that agricultural capital would be the natural allies of the commercial bourgeoisie.

The war years had seen the commercial bourgeoisie rise to a position of dominance, but the landed interests had not been untouched by significant changes, as the pressures for the establishment of capitalist farms grew. On various occasions since the establishment of the Republic, attempts had been made to introduce some kind of land reform, but each attempt had encountered resistance or had been ineffective. Bayar, in 1937, as Prime Minister, had included land reform in his programme, but nothing was achieved. In 1945, another attempt was made, and this time a "Land Law" was passed through the Grand National Assembly, which aimed at the creation of large and efficient capitalist farms, and the abolition of absentee landlordism, although it envisaged small family farms co-existing with capitalist ones. The law aimed to distribute land among the small peasants so that they could use their land for subsistence, but could also work for the capitalist farmers as share-croppers or wage labourers.

The strong opposition to this law was firmly associated with the important names of Bayar and Menderes. Bayar represented the interests of commercial capital as the leader of the Business Bank Group, while Menderes was a big landlord in the fertile plains of Western Anatolia who was rapidly becoming a big capitalist farmer himself. The alliance between Bayar and Menderes indicated the source of the new pressures for liberalisation of the economy. Land reform was the last straw for this class alliance, coming after a haphazard series of RPP policies.

While the RPP was still nominally in control of the state machinery and the pursuit of etatist policies, the conjuncture of the times meant that commercial and agricultural capital together could demand more direct participation in the use of state power. The etatist policies of the RPP and of the Kemalist regime had been successful in the creation of a strong, dominant capitalist class; the monopolistic control of state power by the crystallisation of class structure was at least possible. Ironically, the RPP was not the means by which this alliance could achieve its dominance over the rest of the social relations of production. A new party, the Democrat Party (DP), was formed in 1946 and led by Bayar and Menderes (Erogul, 1970; Agaoglu, 1972).

Inonu, as the leader of the RPP was hostile to the development of the DP, not least because the new leaders of the DP were former longstanding and important members of the RPP. What the remaining leadership of the RPP did not realise was that the RPP had never in its history been the representative of any particular class fraction of fractions, with any consistency. It had always tried to promote capitalist development by giving priority to industry, but at times it had been compelled to support other class fractions in the struggle for dominance. The RPP has proved more successful in reconciling contradictory interests in order to speed development, than in committing itself to an unequivocal association with commercial and agricultural interests. The RPP was still bogged down by explicit etatism in its programme when etatism as the legitimation of the role of the state in capitalist development was no longer required. The birth of the DP in 1946 does not, however, mark the death of etatism. The DP view could loosely be described as "etatism is dead, long live etatism, but keep it quiet". This new formulation of implicit etatism meant that the state was to work for the dominant class fractions, and that bureaucratic power was to curtailed and controlled.

Before they came to power the DP was promoting the view that the interests of the alliance between commercial and agricultural capital could best be served by opening the Turkish economy to world markets. In 1947, a report was prepared under the auspices of the RPP government, but by a group associated with the DP, which came to be called the 1947 development plan. [34] This report was not a development plan as such, but it indicated the priorities and interests of the dominant fractions. It was intended as one means of persuading international capital that Turkey should be part of the newly reconstituted world economic order which was established under the Bretton-Woods agreement of 1944. The draft of this plan was, in effect, an industrialisation programme for Turkey, in accordance with RPP policies. The final document, however, was very different, as it had been changed in response to foreign pressures, leading to the withdrawal of the emphasis on industrialisation, and adding the devaluation of Turkish currency. In the final plan, industrialisation, which had formed the basis for the pre-war policy of import-substitution, had disappeared from the agenda for capitalist development. Transport, communications, agriculture and energy were to be promoted, while the production of consumer goods, including the state sector, was to be transferred to private capital.

This plan never became official RPP policy, but it illustrates the contradictory position of the RPP at this period. While its declared policy was to promote the interests of industry and commerce, the non-correspondence between the political and economic levels, meant that it could not do both. Its political rival the DP, in supporting unconditionally the interests of the commercial bourgeoisie against those of the industrial bourgeoisie, developed a much more realistic policy at this conjuncture. In the view of the DP, the economic role of the state should be confined to the provision of infrastructural services for private capital, but the principles of etatism could be so broadly interpreted, that this view need not have been seen as in contradiction to them. The issue being the differences of policy was not interpretation of etatism, but the struggle of the alliance of commercial and agricultural capital to gain control of state power. In this struggle it was necessary for the DP to be seen to be opposing the etatist policies of the RPP, as the RPP interpretation of etatism represented support for the interests of the industrial bourgeoisie.

In 1950, the DP came to power with a landslide victory which allowed a distinct class alliance to monopolise control of state apparatuses. [35] Merchant and agricultural capital had won the battle for the control of the state, at least temporarily. This decisive victory led to the abrupt rejection of import-substitution as the blue print for Turkey's economic development, in favour of an export-promotion model. This dramatic change in policy was, therefore, the direct outcome of contradictions within Turkish capitalism at this conjuncture. The rise to power of the DP coincided with the economic boom created by the Korean War. This was a period of inflation and of rapid growth in Turkish exports, and in agricultural production. With the tailing off of this boom in 1953-54, Turkey began to face growing problems of competing in the export market, which led fairly quickly to total reversal of the DP's export promotion strategy. A protective trade regime had to be re-introduced, with a quotas and restrictions on imports. This meant that the demand which had been built up for imported manufactured goods for the Turkish economy now had to be met by the indigenous industry. This in turn led to the expansion of Turkish industrial capital, and contributed to the development of the processes of generalised commodity production, and to the expanded reproduction of capital in the further transformation of Turkish capitalism. The new specific contradictions within Turkish capitalism brought about by internal and external developments, brought industrial capital to the forefront of Turkish capitalism yet again. In spite of the DP's declared strategies, and its class basis, the model of development adopted in 1954-55 was to be import-substitution for the second time.

This second period of import-substitution lasted until the crises of the late 1970s, which culminated in the military coup of 1980. [36] The policy changes forced on the DP created new forms of non-correspondence between economic, political and ideological levels. These contradictions ultimately destroyed the DP which was removed from government by a military coup in 1960. This marked the end of the brief period of the monopolistic use of state power by the alliance of commercial and agricultural capital, and prepared the necessary conditions for the rise to dominance of monopoly/industrial and financial capital.

CONCLUSION

Contradictions within Turkish capital and between Turkish and international capital have been, and will continue to be the source of further class conflicts, with the intervention of the military regime, which came into existence after the 1980 military coup, marking another stage in these struggles.

In this chapter, I have laid out conjunctures in the emergence of Turkish capitalism. I have tried to show, in particular, key developments in class structures and the differentiation of the emerging bourgeoisie. I have argued throughout that the structure and organisation of the Turkish state during this period underwent systematic changes, which were consistent with the logic of Turkey's capitalist development. The struggle between the dominant classes over the monopolistic use of state power culminated in the alliance of commercial and agricultural capital during and after the Second World War. The coming to power of this alliance under the umbrella of the DP in 1950, marked a new stage in the transformation of Turkish capitalism. The domain of Turkish politics was opened up for overt class struggle. In the struggles that developed, the different fractions of the dominant classes (monopoly/industrial, financial, commercial and agricultural) were forced to share state power between them. The political system which existed up to September 1980 did not allow the resolution of this crisis of representation in favour of any one fraction of the bourgeoisie. Given the rapid accumulation of industrial and financial capital, these fractions of the bourgeoisie were able to establish their economic dominance, but this dominance was not complemented at the political level because they were forced to share state power with agricultural and commercial fractions of the bourgeoisie. The democratic parliamentary system, therefore, actually obstructed the resolution of the struggle to monopolise state power. It was this continuing inability of monopoly/industrial and financial capital to dominate state power which lies behind the instability of Turkish democracy.

I have also argued that the decision to adopt import-substitution as the key strategy for economic development has had a direct impact on the shaping of state structures. The import-substitution model and the aim of the closed economy were practical policies to employ at this stage in the development of Turkish capitalism, but when these policies reached a stage of being counterproductive they needed to be changed if Turkish capitalism was to advance further. The political, economic and ideological practices, however, which had been founded, were geared to import-substitution and their existence hindered a gradual transformation to an outward-looking economy based on a strategy of export-promotion. I have not undertaken an examination of the difficulties which the import-substitution model of development created nor of the ensuing arguments to switch to export-promotion. These issues are dealt with elsewhere in this book. In this chapter, I have simply tried to analyse critical conjunctures in the emergence of the class power of the bourgeoisie and to indicate that the present crisis is rooted in the organisation of Turkish capitalism before 1950.

NOTES

1. Cengiz Arin and Feroz Ahmad have read the earlier versions of this chapter. I am grateful for all their comments.

2. See Chapter 1 of this book, and also Ramazanoglu, 1980.

3. See Chapter 1 of this book.

4. One of the better explanations of this period is provided by Shaw and Shaw, 1977. To give it its proper name, Tanzimat-i Hayriye or "Auspicious Reorderings" was a period of sustained legislation and reform that modernised Ottoman state and society, contributed to the further centralisation of administration, and brought increased state participation in Ottoman society between 1839 and 1876.

> "...by extending the scope of Ottoman Government far beyond its traditional bounds to include the right and even the duty to regulate all aspects of life and changing the concept of Ottoman reform from the traditional one of attempting to preserve and restore old institutions to a modern one of replacing them with new ones, some imported from the West. The successes as well as the failures of the Tanzimat movement in many ways directly determined the course reform was to take subsequently in the Turkish Republic to the present day." (p.55)

   For another interesting account of this period and its significance in the context of the growth and the decline of the Ottoman Empire and the foundation of the new Turkish Republic, see Yerasimos, 1974.

5. See Findley, 1980, for the nature and the structure of Ottoman bureaucracy, and Shaw and Shaw, 1977, pp.123-128, for the role of ethnic divisions (millet) during the Tanzimat period, and also Yerasimos, 1974, pp. 324-330; also see Sunar, 1974.

6. An Armenian banker from the Galata financial district of Istanbul was appointed to manage the privy purse by Sultan Abdulhamit II (Shaw and Shaw, 1977, p.222); for an examination of the role of the foreigners in controlling the finances of the Ottoman Empire, see Keyder, 1980.

7. For a good historical account of the Young Turk period (1908-18), see Ahmad, 1969, and Shaw and Shaw, 1974, pp.273-339. This period also coincides with attempts to establish a "national economy", see Toprak, 1982.

8. It must be noted that the Young Turks were not aiming at overthrowing the Ottoman dynasty although they saw the Sultan as weak and unable to stand up to the imperialist powers and their agents. Rather, they were aiming at improving the situation by introducing social and political reforms. Mustafa Kemal and his followers, on the other hand, saw the overthrow of the Ottoman dynasty as being essential for the establishment of a totally new order, thus breaking with the past and concentrating on the future. The establishment of the Grand

National Assembly was a first step towards this purpose which was finally realised with the proclamation of the Republic in 1923. The Sultan had escaped on a British warship in 1922.

9. Two views were current, of which the first advocated a return to the rule of the Sultanate, and the second advocated the formation of a republic, and a complete break with the legacy of the Ottoman Empire. Mustafa Kemal criticised the first view and urged full support for the second in one of his speeches. Mustafa Kemal Ataturk, 1964.

10. For a classic work on the examination of the application of the import-substitution model see Hirschman, 1968. A recent book by Bienefeld and Godfrey, 1982, is also a useful source to refer to. In the case of Turkey, the World Bank Report on the Turkish economy, 1975 and 1982; the books by Krueger, 1974; Hershlag, 1968 and Walstedt, 1980 are essential sources to consult.

11. See Avcioglu, 1968, p.165.

12. Although the Congress did adopt a resolution on the recognition of trade unions, no action was taken. Okcun, 1970.

13. See Okcun, 1970, pp.255-256.

14. Celal Bayar, who later became a leader of the Democrat Party, was appointed as the director of the Business Bank (Is Bankasi). The Board of Governors of the bank was dominated by merchants who were to be known as "The Business Bank Group" and who played an important role in the development of Turkish capitalism. Avcioglu, 1968, pp.180-182.

15. See Shaw and Shaw, 1977, p.388.

16. Given the composition of "The Business Bank Group", one can assume that the Group had no real incentive to support industry when the conditions were so suitable for international trade, which would benefit the interests that the Group represented.

17. The world recession in 1929 contributed greatly to the decline of the export trade, thus causing severe shortages of foreign exchange, and eventually inability to import even the most basic goods, which meant a rapid industrialisation programme. See Tekeli and Ilkin, 1977.

18. There seems to be a general agreement in the literature on this point, see Gulalp, 1980, p.45, but I see this generally accepted view that the 1929 recession gave rise to etatism as being rather simplistic. My interpretation is different, as I try to argue here.

19. See Boratav, 1974, p.104.

20. A detailed account of this legislation can be found in Boratav, 1974, pp.125-126.

21. See Boratav, 1974, pp.128-129.

22. Soviet planning experience influenced the decision to industrialise within the framework of a planned economy. A Soviet delegation, led by Prof. Orlov, contributed to the preparations of the First Five-Year Plan. They also supported the decisions in favour of industrialisation taken at the Izmir Economic Congress, which recommended priority to be given to the development of domestic industry, and the use of immediately available raw materials. Tayanc, 1973, pp.98-99, and Tekeli and Ilkin, 1982.

23. The fears of increased state intervention and of collectivism are dealt with in Boratav, 1974, pp.167-170; Goymen, 1976, pp.102-105 and Tekeli and Ilkin, 1982.

24. Adnan Menderes, who with Bayar later founded the Democrat Party, was one of the leading members of the Free Party, advocating liberal trade regimes. It would be misleading to argue that the Free Party was formed to oppose Ataturk's Republican People's Party, as it was not only directly opposed to Ataturk and his regime, but it was in fact used by Ataturk the strength of the opposition. For an interesting if not altogether clear account, see Kucuk, 1978, pp.106-116.

25. The response to the formation of the Free Party was so strong that Ataturk saw it as a threat to his personal authority, and closed it.

26. _Kadro_ opened its pages to committed exponents of _etatist_ policies. In 1933, the Prime Minister Inonu attacked the opponents of _etatism_ as naive, and argued that _etatism_ was a means of self defence, see _Kadro_, October, 1933, cited in Goymen, 1976, p.99. A facsimile edition of the full set of _Kadro_ was recently prepared by Alpar and published in three volumes, 1980.

27. See Boratav, 1974, pp.183-184. Bayar is generally considered to have played a vital role in slowing down the implementation of _etatism_ after he became Prime Minister in 1937. Bayar should not be seen as opposed to _etatism_, however, but rather as adapting the interpretation of _etatism_ in order to meet the demands of private capital. The public sector was to concentrate its activities on the development of heavy industry, as envisaged by the Second Five-Year Plan of 1938, while the production of consumer goods was largely to be left to the private sector. The Second Five-Year Plan was not implemented because of the outbreak of the Second World War.

28. For an explanation of the ideological practices implemented by the _Kemalist_ regime. See Keyder, 1979, pp.14-15 and also Ozbudun, 1981.

29. This decision was not only the direct outcome of internal contradictions between industry and agriculture, but also of contradictions arising from the international division of labour.

30. As commodity production becomes increasingly generalised and expands into agriculture, the differentiation of the peasantry also increases, and this has some serious consequences for the rural sector: a) the dependence of peasants on their landlords increases, b) small direct production no longer remains domestic based, but instead is largely influenced by the production relations of the expanding exchange economy, which is increasingly determined by industrial and

financial capital as the expanded reproduction of capital becomes the dominant form of production, and c) there is a growing trend for the small direct producers being transformed into wage earners, either as agricultural or, especially, as industrial proletariat and are forced to migrate to cities.

31. With the legislation of 1936 in particular, the conditions in the labour market became very oppressive, see Avcioglu, 1968, pp.222-224. This legislation was symptomatic of increasing repression and dwindling popular support.

32. This law was extremely detailed and comprehensive. Avcioglu, 1968, pp.221-222, deals with some of the important points, and a full version can be found in Boratav, 1974, pp.326-335.

33. See Okte, 1951, cited in Avcioglu, 1974 p.226.

34. See Tekeli and Ilkin, 1974. Gulalp sees this plan as the main document which shaped the course of economic policy after the war, Gulalp, 1980, p.48.

35. For an examination of the role of the rural classes in the downfall of the Republican People's Party, see Keyder, 1979, pp.17-20, and for a detailed analysis of the political system of that period, see Karpat, 1959.

36. For further analysis of the period from 1950 to the present day, see Ahmad, 1977; Ahmad, 1981; Keyder, 1979; Pamuk, 1981 and Kongar, 1976. See also Krueger, 1974, on post-1950 foreign economic relations, and Hershlag, 1968, for a general introduction to the development of an inward-looking Turkish economy.

REFERENCES

Agaoglu, Samet (1972); Demokrat Partinin Dogus ve Yukselis Sebepleri (Reasons for the Emergence and the Rise of the Democrat Party), (publisher's name not available),

Ahmad, Feroz (1969); The Young Turks, Oxford University Press, Oxford

Ahmad, Feroz (1977); The Turkish Experiment in Democracy, 1950-1975, Hurst, London

Ahmad, Feroz (1980); "Vanguard of a Nascent Bourgeoisie: The Social and Economic Policy of the Young Turks, 1908-1918" in O. Okyar and H. Inalcik (eds.), Social and Economic History of Turkey, Meteksan, Ankara

Ahmad, Feroz (1981); "Military Intervention and the Crisis in Turkey", MERIP Reports, No.93, pp.5-24

Ahmad, Feroz (1981); "The Political Economy of Kemalism" in A. Kazancigil and E. Ozbudun (eds.), Ataturk, Founder of a Modern State, Hurst, London

Alpar, Cem (1980); Kadro (Cadre), Vols.1-3, Ankara Iktisadi ve Ticari Ilimler Akademisi, Ankara

Ataturk, Mustafa Kemal (1964); Soylev (Discourse), Vol.1, Ankara

Avcioglu, Dogan (1968); Turkiye'nin Duzeni (The Turkish Social Order), Bilgi Yayinevi, Istanbul

Aydemir, Sevket.S. (1968); Devrim ve Kadro (Revolution and Cadre), Bilgi Yayinevi, Istanbul

Bienefeld, Manfred and Godfrey, Martin (eds) (1982); The Struggle for Development: National Strategies in an International Context, Wiley, New York

Bila, Hikmet (1979); CHP Tarihi 1919-1979 (The History of the RPP, 1919-1979), Doruk Matbaacilik Sanayii, Ankara

Birtek, Faruk and Keyder, Keyder (1975); "Agriculture and the State: An Inquiry into Agricultural Differentiation and Political Alliances; The Case of Turkey", Journal of Peasant Studies, Vol.2, No.4, pp.446-67

Boratav, Korkut (1974); Turkiye'de Devletcilik (Etatism in Turkey), Gercek Yayinevi, Istanbul

Bulutoglu, Kenan (1970); Turkiye'de Yabanci Sermaye (Foreign Capital in Turkey), Gercek Yayinevi, Istanbul

Erogul, Cem (1970); Demorat Parti: Tarihi ve Ideolojisi (The Democrat Party: Its History and Ideology), Faculty of Political Science, Ankara

Findley, Carter V. (1980); Bureaucratic Reform in the Ottoman Empire: The Sublime Porte 1789-1922, Princeton University Press, Princeton

Goymen, Korel (1976); "Stages of Etatist Development in Turkey: The Interaction of Single Party Politics and Economic Policy in the Etatist Decade, 1930-1939", Studies in Development, No.10, pp.89-113

Goymen, Korel and Tuzun, Gurel (1976); "Foreign Private Capital in Turkey: An Analysis of Capital Imported Under the Encouragement Law 6224", Studies in Development, No.11, pp.55-79

Gulalp, Haldun (1980); "Turkiye'de Ithal Ikamesi Bunalimi ve Disa Acilma" (The Crisis of Import-Substitution in Turkey and Outward-Orientation), Studies in Development, Vol.7, Nos.1-2, pp.37-65

Hershlag, Z.Y (1968); Turkey: The Challenge of Growth, Brill, Leiden

Hirschman, A.O. (1968); "The Political Economy of Import-Substituting Industrialisation in Latin America", The Quarterly Journal of Economics, Vol.82, February

Issawi, Charles (1980); The Economic History of Turkey 1800-1914, Chicago University Press, Chicago

Karal, Enver (1981); "The Principles of Kemalism", in A. Kazancigil and E. Ozbudun (eds.), Ataturk, Founder of a Modern State, Hurst, London

Karpat, Kemal (1959); Turkey's Politics: The Transition to a Multi-Party System, Princeton University Press, Princeton

Kazancigil, A. and Ozbudun, E. (eds.) (1981); Ataturk, Founder of a Modern State, Hurst, London

Kazancigil, Ali (1981); "The Ottoman-Turkish State and Kemalism", in A. Kazancigil and E. Ozbudun (eds.), Ataturk, Founder of a Modern State, Hurst, London

Keyder, Caglar (1979); "The Political Economy of Turkish Democracy", New Left Review, No.115, pp.3-44

Keyder, Caglar (1980); "Ottoman Economy and Finances (1881-1918)" in O. Okyar and H. Inalcik (eds.), Social and Economic History of Turkey, Meteksan, Ankara

Kili, Suna (1976); 1960-1975 Doneminde Cumhuriyet Halk Partisinde Gelismeler: Siyaset Bilimi Acisindan Bir Inceleme (Developments in the Republican People's Party during the Period of 1960-1975: An Analysis from the Political Science Perspective), Bogazici Universitesi Yayinlari, Istanbul

Kongar, Emre (1976); Imparatorluktan Gunumuze Turkiye'nin Toplumsal Yapisi (The Social Structure of Turkey from Empire to Date), Cem Yayinevi, Istanbul

Krueger, Anne O. (1974); Foreign Trade Regimes & Economic Development: Turkey, National Bureau of Economic Research, Columbia University Press, New York

Kucuk, Yalcin (1978); Turkiye Uzerine Tezler (Theses on Turkey), Vol.1, Tekin Yayinevi, Istanbul

Kushner, David (1977); The Rise of Turkish Nationalism 1876-1908, Frank Cass, London

Landau, Jacob (1981); Pan-Turkism in Turkey, Hurst, London

Meray, Seha (1970, 1972, 1973); Lozan Baris Konferansi: Tutanaklar - Belgeler (The Lausanne Peace Treaty: Proceedings and Documents), Vols.1-5, Faculty of Political Science, Ankara

Okcun, Gunduz (1970); Turkiye Iktisat Kongresi, 1923 - Izmir: Haberler, Belgeler, Yorumlar (Economic Congress of Turkey, 1923 - Izmir: Reports, Documents, Commentaries), Faculty of Political Science, Ankara

Okte, A.Faik (1951); Varlik Vergisi Faciasi (The Tragedy of the Wealth Tax), (publisher' name not available)

Okyar, Osman (1975); Turkiye Iktisat Tarihi Seminari (Seminar on Turkish Economic History), Hacettepe Universitesi Yayinlari, Ankara

Okyar, Osman and Inalcik, Halis (eds.) (1980); Social and Economic History of Turkey, Meteksan, London

Ozbudun, Ergun (1981); "The Nature of the Kemalist Political Regime" in A. Kazancigil and E. Ozbudun (eds.), Ataturk, Founder of a Modern State, Hurst, London

Pamuk, Sevket (1981): "The Political Economy of Industrialisation in Turkey", MERIP Reports, No:93, pp.26-32

Ramazanoglu, Huseyin (1980); "After Dependency: A Contribution to the Debate", Journal of Area Studies, No.2, pp.22-25

Shaw, Stanford and Shaw, Ezel (1977); History of the Ottoman Empire and Modern Turkey,, Vol.2, Cambridge University Press, Cambridge

Silier, Oya (1975); "1920'lerde Turkiye'de Milli Bankaciligin Genel Gorunumu", (The General Outlook of National Banking in Turkey in 1920s) in O. Okyar, Turkiye Iktisat Tarihi Seminari (Seminar on Turkish Economic History), Hacettepe Universitesi Yayinlari, Ankara

Sunar, Ilkay (1974); State and Society in the Politics of Turkey's Development, Ankara University, Faculty of Political Science, Ankara

Tayanc, Tunc (1978); Sanayilesme Surecinde Elli Yil (Fifty Years in the Process of Industrialisation), Milliyet Yayinlari, Istanbul

Tekeli, Ilhan and Ilkin, Selim (1974); Savas Sonrasi Ortaminda 1947 Turkiye Iktisadi Kalkinma Plani (The 1947 Turkish Economic Development Plan in the Aftermath of the War), Middle East Technical University, Ankara

Tekeli, Ilhan and Ilkin, Selim (1977); 1929 Dunya Buhraninda Turkiye'nin Iktisadi Politika Arayislari (Turkey's Search for Economic Policies during the 1929 World Depression), Middle East Technical University, Ankara

Tekeli, Ilhan and Ilkin, Selim (1982); Uygulamaya Gecerken Turkiye'de Devletciligin Olusumu (The Emergence and the Implementation of Etatism in Turkey), Middle East Technical University, Ankara

Toprak, Zafer (1982); Turkiye'de "Milli Iktisat", 1908-1918 ("National Economy" in Turkey, 1908-1918), Yurt Yayinlari, Ankara

Tuncay, Mete (1981); Turkiye Cumhuriyeti'nde Tek-Parti Yonetiminin Kurulmasi (1923-1931) (The Establishment of Single Party Rule in the Turkish Republic, 1923-1931), Yurt Yayinlari, Ankara

Uras, Gungor (1979); Turkiye'de Yabanci Sermaye Yatirimlari (Foreign Capital Investments in Turkey), Iktisat Yayinlari, Istanbul

Walstedt, Bertil (1980); State Manufacturing Enterprise in a Mixed Economy, World Bank/Johns Hopkins University Press, Baltimore

World Bank (1975); Turkey: Prospects and Problems of an Expanding Economy, World Bank, Washington D.C.

World Bank (1982); Turkey: Industrialisation and Trade Strategy, Vol.II, World Bank, Washington D.C.

Yerasimos, Stefanos (1974); Azgelismislik Surecinde Turkiye: Bizanstan 1971'e (Turkey in the Process of Underdevelopment: From Byzantium to 1971), Gozlem Yayinlari, Istanbul; Turkish translation from the French original.

3 The politics of industrialisation in a closed economy and the IMF intervention of 1979

HUSEYIN RAMAZANOGLU

Turkey today presents a picture of a social formation whose political future looks increasingly uncertain, while its credibility in world markets is steadily improving. The chief problem facing the present regime and the Turkish ruling classes in directing the transformation of the economy, arises from the nature of the contradictions of Turkish capitalism. These contradictions are very sharp, and with the accelerated development of capitalism in Turkey, they will intensify and there are likely to be periodic interruptions rather than a smooth process of transformation. The present regime and its successors may find the volatility and strength of these latent forces difficult to control. These problems are obviously not new, and in order to understand how the contradictions of Turkish capitalism have developed since the 1950s, and how they culminated in the military intervention of 1980, we need to look at the origins and the nature of these problems, and at the impact of the IMF intervention which finally paved the way for the introduction of policies aimed at radically transforming Turkish capitalism.

OBSTACLES TO THE EXPANSION OF TURKISH CAPITALISM: 1923-1980

From the establishment of the Republic in 1923, Turkey pursued a policy of industrialisation within a closed economy. [1] The key development strategy was for the state to intervene in the economic sphere, in order to lay the foundations for the development of private enterprise and also to stimulate growth in those sectors of the economy neglected by private capital. The size of the domestic market and general international acceptance of this _etatist_ strategy made it possible for successive Turkish governments to carry on building the economy behind closed doors, thus contributing to the emergence of a strong national bourgeoisie. The industrialisation strategy adopted was that of import-substitution, a policy which was aimed at discouraging reliance on foreign capital and actively encouraging the accumulation of indigenous capital. This general strategy continued somewhat variably, but on the whole successfully, until the late 1960s, when the expansion of indigenous capital began to reach its limits within the framework of the domestic market. The need to find new fields for investment and new markets for Turkish goods necessitated a reappraisal of the inward-looking economic strategy. This pressure for change was also affected by the growing crisis of the world economy in the 1970s. Centres of international capital had been financing Turkey's growing trade deficit which had been building up during the years of the closed economy, but with the onset of the world recession they were no longer willing to do so. This was, in part, because of crises in the Western economies, and also because it was thought that the Turkish economy had developed sufficiently to be able to attract international capital without outside help.

By the end of 1980, Turkey's total foreign debts amounted to $17.8 billion. [2] These debts did not make Turkey a very desirable location for further credits because of the risks that loans and credits could not be recovered in the foreseeable future. Turkey's ability to service its debts let alone repay them began to lessen as new developments adversely effected the performance of the economy. It was recognised that immediate economic strategies were needed to deal with this rapidly worsening situation, but the implementation of these strategies were obstructed by political considerations. This political blocking of economic change brought Turkey's credit worthiness to a very low point indeed. The fact remained, nevertheless, that if the political obstacles to economic change could be removed, other conditions were suitable for the transformation of Turkish capitalism. Turkey was potentially an important market, and one which could be fairly rapidly developed. The domestic market had been gradually built up over previous decades when it was shielded from the rigors of international competition, and industrial capital had been the main beneficiary of import-substitution and a protective trade regime. As a result of these policies the market had been flooded with inferior quality and highly priced manufactured goods, and was eventually near saturation point. A combination of internal and external factors had, however, put the industrial sector under tremendous pressure, thus making the accumulation of capital in the industrial sector increasingly difficult.

The balance of payments problem, in particular, had become so severe, that from the mid-1970s onwards the manufacturing sector was forced to cut output because of the lack of foreign exchange to finance imports. Worsening market conditions did not help the situation and, in many cases, manufacturers were forced to operate at only a third of available capacity. This led the industrial bourgeoisie to look outside Turkey for new sources of capital investment and joint ventures. This meant opening the economy to foreign capital, and consequently accepting the threat which would come from the penetration of the domestic market by high quality products of foreign origin which were "good value for money". The sheer size of the potential market for relatively cheaper and better quality products was certainly too attractive for international capital to ignore. Turkey's strategic location also provided access to the lucrative Middle East markets which were rapidly opening to international capital. The industrial and financial bourgeoisie, therefore, had to reconsider where their best interests lay, once the highly protected closed economy ceased to reinforce the economic basis of their dominance.

The absence of efficient and controlled money, banking and credit markets in the economy meant that the growing needs of monopoly/industrial and financial capital could not be met. Apart from the political considerations which blocked the opening of the economy, there were no channels to international money markets, which was another obstacle to the investigation of a desperately needed and long overdue export-promotion drive. The famous etatist legislation, the "Law for the Protection of the Value of Turkish Currency", passed in 1930, had effectively blocked the establishment of such channels by making the circulation of Turkish currency in international markets impossible. There was, therefore, no pressure to develop structures in the economy which could cope with the real and growing needs of Turkish capitalism, when these needs were thwarted by the juridico-political structures of the state. Various attempts had been made to change not only this law but also others, but the implementation of these legislative changes was obstructed by the bureaucracy, which came to treat these changes as anomalies that were not to be taken seriously. The introduction of a system aimed at the radical transformation of Turkish capitalism would immediately come up against the serious problem of controlling the bureaucracy and getting its co-operation in implementing new policies. The dwindling of remittances coming from migrant workers in Europe from the mid-1970s onwards formed yet another bottleneck in the economy. [3] The policies adopted by successive governments to attract and stimulate new investments in the economy by migrant workers had been somewhat haphazard and inefficient. Enormous funds accumulated by Turkish workers in Europe since the early 1960s stayed largely in the form of deposits locked up in the vaults of European banks. It is estimated that this sum is around $6 billion, and only a very small proportion of this money trickles back to Turkey. These remittances, despite their insufficiency, had propped up the economy during the crisis which followed the increase in oil prices in 1973, but they gradually lost their significance as new emigration ceased, and the problems of the economy became too severe to be solved by such palliative measures.

Another major obstacle to the expansion of Turkish capitalism arose from the position of the landed interests and the importance of the agricultural sector in Turkey. Because of their political power, the landed interests had secured very effective channels of representation in the parliamentary system. Hence, they had managed to prevent several, albeit half-hearted, attempts to implement land reforms in the 1960s and 1970s, despite the fact that legislation had been passed through parliament establishing a pilot project in Urfa, a Southeast province of Turkey, as a first step towards general land reform. The gradual reallocation of resources in the economy in favour of industry and the capitalisation of agriculture had not been sufficiently rapid or effective to improve the quality of agricultural production or, more importantly,to bring the landed interests and the agricultural sector under the domination of monopoly/industrial and finance capital and the market economy. The full potential of the agricultural sector could only be realised if these two conditions could be met successfully, thus turning Turkey into the bread basket of Europe and the Middle East.

The 1970s, therefore, became a turning point in the development of Turkish capitalism, when import-substitution and other inward-looking economic strategies reached the limits of their usefulness. The pride and joy of etatism, the State Economic Enterprises, had become the lame ducks of the Turkish economy and were in need of radical restructuring or dismantling, but because of their political significance they had been left more or less intact since their inception. [4] During the long period of import-substitution and state intervention in economic development, however, Turkish capitalism had been developing, and at the same time creating the conditions which would allow considerable potential for further development. [5] Through the reallocation of resources within the international division of labour, Turkey could now be developed as a market for investment in capital and semi-capital intensive technologies used in the production of electrical and non-electrical machinery; also for machine-tool, electronics, pharmaceuticals, durable consumer goods, agricultural products and construction. Within the restructuring of the international division of labour, which is now taking place, Turkey had the potential to develop a role as a supplier of such commodities. The achievement of such rapid changes in patterns of production is, however, dependent upon rapid change in overall economic strategy. Today, Turkey is having to undergo a drastic departure from a relatively independent and inward-looking set of policies, to the acceptance of outward-looking policies geared for export-promotion and the attraction of foreign capital. These economic changes cannot take place without a widespread political and social transformation and it is only to be expected that Turkish capitalism is going through a period of considerable upheaval and readjustment.

The new economic strategies which are now being put into effect are aimed at attracting a steady flow of foreign capital for investment in the Turkish economy (Bulutoglu, 1970; Goymen and Tuzun, 1976; Uras, 1979). It is widely acknowledged, however, that an influx of foreign capital will mean an increasing loss of control over those sectors of the economy which traditionally have been controlled by Turkish capital. It is also acknowledged that an influx of foreign capital is not likely to take place immediately. There will have to be a transitional period during which production is restructured and money markets are established, which will tie Turkey more closely to centres of international capital. The other very important condition, which will have to be met, is that of political stability. The transformation of the economy will undoubtedly bring about fundamental and irreversible changes in the whole structure of society. Turkey's economy will become much more tightly integrated into the world capitalist system, and its self-imposed barriers to international trade and investment will have to be remodelled as an outward-looking economy, along the lines of e.g. Korea, Brazil, Mexico.

This process of transformation seems to be what is intended by the various centres of international capital and also by the dominant fraction of the Turkish bourgeoisie, i.e. monopoly/industrial and financial capital (TUSIAD, 1980-81; Balassa, 1981a). The only problem, of course, is that Turkey's economic and political structures have been geared to the maintenance and the reproduction of a closed economy since the early days of the republic. The switch from import-substitution, which has been the main strategy for a closed economy, to export-promotion, which is advocated as the main strategy for an open economy, presents serious political problems, as well as economic ones. The necessary changes cannot be implemented in a relatively short space of time because they entail a major restructuring of the state apparatuses and the transformation of all levels of the Turkish social formation.

The gradual accumulation of the economic and problems, which preceded the 1980 coup, was generally recognised by the dominant fractions of the Turkish bourgeoisie. These problems had been foreseen, but attempts either to prevent them from occurring or to resolve them were hampered by the juridico-political structures of the Turkish state whose roots were firmly embedded in etatism.

A major casualty of the need to open the economy has been the Turkish parliamentary system. Liberal democracy which had been developed since the foundation of the Republic had outlived its usefulness to dominant fractions of capital. By virtue of its structure and nature the parliamentary system had fuelled the sharpening of contradictions between competing fractions of the Turkish bourgeoisie rather than providing an arena within which their differences might be resolved. It had provided a necessary political shell for the development of Turkish capitalism based on the continued implementation of etatist principles, but had failed to provide a political framework which was suitable for an effective transformation of Turkish capitalism. To appreciate the inevitable dissolution of the parliamentary system in Turkey, and its replacement with political structures consistent with the needs of an open economy, it is necessary to look briefly at the development of the struggle between fractions of the bourgeoisie that occurred within the framework of this parliamentary system.

THE POLITICS OF INDUSTRIALISATION IN A CLOSED ECONOMY

Since the formation of the Turkish Republic, the maintenance of formal political institutions has always been of great importance, and for most of its short history, Turkey has had a democratic political system. This appearance of democracy, however, has not always reflected the political struggle for the control of the state. The real struggle has taken place outside the democratic political processes, but in the struggle, democratic institutions have been utilised to provide a framework of legitimacy. (One might argue that such practices are not unknown in many advanced capitalist countries where "democracy" is not as directly threatened as it is in, so called, less developed countries.)

In the relative safety of a closed economy and with a programme of import-substitution, industrial production for the domestic market grew rapidly. During the 1960s, Turkey had an annual economic growth rate of 7-8 per cent which was one of the highest in the newly industrialising countries in the world. The manufacturing sector became well established in the economy with healthy returns on investment. During this period, large scale production began to develop, changing the character of Turkey's industry which had previously been dominated by numerous small units. In the small towns of Anatolia, many small local firms became sub-contractors to larger firms, or went out of business altogether. Although workshop production remained an important sector, in a relatively short time, the economy became dominated by large monopolies and holding companies. Some firms tried to break into markets which had potential for expansion and managed to become suppliers to major sectors of the economy, but many could not survive increasingly adverse market conditions, or grew at a much slower rate than before.

Another major development on the economic front was the formation of new banks or the take-over of existing banks by the major holding companies. With finance provided internally, the holdings could expand more extensively. Competition for the control of sources of finance became intensified and was further compounded by personality clashes between big businessmen, as for example, the rivalry between the Koc and Sabanci families who between them owned and controlled two of the largest enterprises in the country.

The economic power enjoyed by the monopoly/industrial and financial bourgeoisie was indisputable, but they had to share political power with the other fractions of the bourgeoisie, namely the commercial and landed interests, since they had failed to dominate these fractions. This situation had inherent difficulties which actively hindered the further transformation of Turkish capitalism.

The complexities of party politics in Turkey during this period may seem somewhat technical or even parochial to an outsider, but the general significance of the instability of Turkey's democracy can be clarified by seeing the machinations of individual politicians as elements in a fiercely fought class struggle, whose goal was control of the state. During the 1960s and 1970s the fragmentation of the Turkish bourgeoisie became increasingly crystallised, as a consequence of the uneven development of capitalism, and political parties formed, changed and made alliances as they variously represented different interests in this uneven development.

The Justice Party (JP), which was formed after the 1960 coup, was identified right from the beginning with the interests of monopoly/industrial and financial capital (Weiker, 1963; Harris, 1970; Trimberger, 1978; Keyder, 1979; Ahmad, 1981). When it became clear that Demirel, the leader of the party, was not prepared to make concessions to other fractions of the bourgeoisie with different interests, splinter parties were formed, such as the Democratic Party, the National Salvation Party (NSP), and the Reliance Party, which represented the various interests of the small town petty-bourgeoisie, absentee landlords and the rentier classes. One other party which also emerged in this period was the neo-fascist Nationalist Action Party (NAP) led by the retired colonel Turkes (Turkes, 1977; Landau, 1981). Turkes was a member of the 1960 Junta, who had fallen out with his colleagues. This fragmentation of the Turkish right meant that the JP could not translate the fairly coherent economic power of its class base into political practices aimed at gaining the monopolistic use of the state power by the monopoly/industrial and financial capital, even with the use of the Parliament which was set up after the 1960 coup in order to achieve political stability in this way. The Republican People's Party (RPP), which was now in opposition, tried to benefit from the fragmentation of the right by manipulating the weaknesses of the Demirel regime.

The growth of industry led to other changes in the balance of political power and parliamentary representation, in response to the growing proletarianisation of the masses. Industry created a demand for wage labour which attracted labour from rural areas to industrial centres. Labour migration started with movements of people to towns and cities inland, but later accelerated into a massive shift of labour into urban areas and also to Europe, particularly West Germany. The liberal constitution of 1961, which laid down new structures and processes to accommodate future developments in Turkish capitalism, also permitted the emergence of a legal socialist party, the Workers' Party of Turkey (WPT). The arbitrary and disorganised nature of capitalist development in this era alienated many workers and intellectuals who were attracted to the WPT. In the general election of 1965, the WPT won 15 parliamentary seats despite widespread harassment, intimidation and violence. This was an important development in Turkish political life because, until this election, the ruling classes, although divided among themselves, had enjoyed a clear run of the field. The basic assumptions of the political system had never been challenged, but now there was a legal and legitimate political party which could benefit from the fruits of bourgeois democracy and at the same time challenge the very foundations of Turkish capitalism.

The working class became highly politicised during the 1960s and workers were organised into two main trade union confederations, Turk-Is and later DISK. Turk-Is was generally considered to be working, by and large, in the interests of the employers, while DISK was a progressive or even revolutionary force which threatened to become the dominant working class organisation. The JP government, and business circles generally, were duly worried by such a development and were anxious to curb the growth of DISK. They wanted to amend the trade union legislation, which was originally introduced in 1963 by Ecevit when he was the minister of labour in the RPP government from 1961 to 1965, in such a way that de facto workers would have to join the unions belonging to Turk-Is. The anger and frustration caused by this move was so great and spontaneous that on 15-16th June, 1970, in response to an appeal from DISK leaders, most workers in the Istanbul industrial belt went on strike (Samim, 1981). This incident was seen as the first step towards revolution and the government itself were so alarmed by this spontaneous show of strength that martial law was declared in the industrial provinces of Turkey.

This decision indicated the weakness of the state once more. The left in Turkey was not sufficiently organised to present a real threat to the state, but the internal contradictions of the Turkish bourgeoisie were so sharp and intractable that the threat was seen as real. Since the democratic parliamentary system enforced some constraints on political action, an alleged threat of communism was used to bring in the army for a second time on 12th March, 1971, in order to provide the conditions for the resolution of these contradictions. [6]

On this occasion a formula was used which differed significantly form that of the 1960 experiment. The army did not stage any military operations to overthrow the government. They invoked the state of martial law, which was already in force, to support their ultimatum, signed by chief of general staff and the heads of the armed forces. This ultimatum demanded that the Demirel government should step down because of its inability to control the apparent breakdown of law and order and its failure to eliminate the dangers posed to the Kemalist regime. The outcome of this coup was the implementation of punitive measures against the left which was seen to be the destabilising force in the country. The WPT was promptly dissolved as were numerous other leftwing organisations. Liberal and leftwing groups and individuals were quite brutally and indiscriminately repressed. A developing contradiction in Turkish capitalism was thus resolved, at least temporarily.

The successive military/civilian governments during the period of military dictatorship from March 1971 to October 1973 tried to introduce legal and political 'changes which would secure the conditions of existence for the dominance of monopoly/industrial and financial capital. These were clearly attempts to resolve the contradiction between economic and political power experienced by these dominant fractions of the bourgeoisie, but the measures taken during this period could not ultimately alleviate this very important contradiction in Turkish capitalism. Various attempts were made to resolve growing economic problems, but they also failed to produce results. The problem was that Turkish capitalism had grown too fast, too soon, and had failed to develop the necessary outlets for further expansion (World Bank, 1975).

The domestic market had been developed as far as was possible, under the conditions of a closed economy. Further economic development could only come from opening the economy to world markets, thus generating new channels for growth and expansion. The bottlenecks in the economy, which have been discussed above, were too constricted to be widened by the normal operation of market mechanisms, so that drastic measures had to be taken to restructure Turkish capitalism. Economic change, however, was hampered by a political framework which did not encourage the required changes. Class struggles between rival fractions of the Turkish bourgeoisie could be confined to a parliamentary form, but this did not provide the necessary scope for restructuring the economy. The crux of the problem was that the uneven development of capitalism had given undisputed economic power to the dominant fractions of the bourgeoisie, i.e. monopoly/industrial and financial capital, but this power was being challenged, and challenged effectively, by those who were not going to benefit from an outward-looking economy. The parliamentary system, thus, became the most important factor for the perpetuation of this non-correspondence between the economic and the political levels. Demirel found himself caught between promoting the interests of monopoly/industrial and financial capital and retaining his electoral base, of landed interests and small businesses, where capitalist production was much less highly developed.

In the meantime, big business had founded an influential body, the Association of Turkish Industrialists and Businessmen (TUSIAD). The primary aim of this association was to act as a pressure group on behalf of monopoly/industrial and financial capital. This meant not only maintaining a close relationship with the JP, but also establishing links with the NAP and NSP, which represented rightwing extremism and Moslem traditionalism in the petty-bourgeoisie. [7]

The various rightwing political parties, however, represented not only uneven development but also different personal ambitions and fractional interests and proved unable to come together in a united front under Demirel's leadership. This disunity gave Ecevit a chance to bring the RPP back to power. Ecevit had become the leader of the RPP by ousting the aging Inonu in 1972. A self-declared social democrat, Ecevit set out to reshape the traditionally conservative RPP into a social democrat party with a broad class base and a populist programme which, he thought, would suit the changing conditions of the development of Turkish capitalism much better. When elections were held in 1973, after a period of rather unsuccessful military dictatorship, the RPP won a third of the votes, but this was insufficient to form a majority government. In order to form a government the RPP had to enter into a coalition with one of the smaller parties and an alliance was accordingly made with Erbakan, the opportunist and capricious leader of the NSP (Ahmad, 1981).

This uneasy alliance, based on opposing class interests, resulted in a weak and conflict-ridden coalition government led by Ecevit. This situation did not meet the dominant fractions' desire for a rightwing government which would restructure political institutions and processes in accordance with the needs of an open economy. Objectively, however, the way in which the classes had already developed in Turkey by this time, and the degree of political representation which existed, made the formation of a strong rightwing government, as desired by these dominant interests, impossible within the existing framework of parliamentary democracy.

One section of these dominant interests, known as the Istanbul Group, under the leadership of Ertugrul Soysal, was well aware of the contradictions and the complexity of Turkish capitalism and did not reject out of hand the social democratic alternative. This was, however, a minority view, and the issue was seen in much more black and white terms by the majority of these interests. The argument within the TUSIAD intensified at this time, as divisions within the power block became increasingly apparent. Ecevit's government, in spite of its weaknesses, served to polarise positions in relation to the desirability of retaining parliamentary democracy.

THE SHARPENING OF CONTRADICTIONS AND THE EROSION OF PARLIAMENTARY DEMOCRACY

The distrust of Ecevit's coalition government expressed by the dominant fractions of the bourgeoisie was further strengthened by the populist rhetoric of the RPP. There was evident support for Ecevit coming from the rapidly expanding working classes and more specifically from left-wing organisations; support which was interpreted as the first step to communism. Populist slogans and banners, new evidence of a politicised working class, frightened the dominant interests and increased their intransigence. The RPP's electoral programme conveyed a hardening of attitudes against monopoly capitalism and the encouragement of a so-called "People's Sector". It was this idea, however unrealistic it might have been, which was seen as a direct threat to the interests of monopoly capital.

Social democracy was now firmly identified with the enhancement of the interests of labour and the protection of the working class and, thus, with the downfall of dominant monopoly/industrial and financial interests (Bulutoglu, 1980). These perceptions of the alleged political power wielded by Ecevit's coalition were in fact far from the truth. As a general rule, although social democracy broadens access to state power, it is, nevertheless, part of the process by which the capitalist system is maintained and reproduced, and by which contradictions are contained. I would argue that the ruling classes in Turkey were, by and large, aware of this fact but chose to manipulate and exaggerate fears of revolution in a campaign to erode democratic processes and to undermine the legitimacy and effectiveness of parliamentary institutions which they did not consider appropriate for swift restructuring of Turkish capitalism. Many observers felt that the dominant interests were being too conservative and short-sighted in their attitude towards Ecevit and the RPP. This may have been true in part, but the fact still remains that a rapid transformation of Turkish capitalism through democratic channels was simply not feasible. When one looks at developments from 1973 to september 1980, one can see that the deliberate and insidious breakdown of Turkish democracy was instigated through the political parties on which democracy was based. [8]

The hostile attitude of the dominant classes towards the parliamentary process in general, and Ecevit's government in particular, was reflected in their decision to stop investment in the economy until radical changes had been made. They did not feel that enough was being done to prepare the economy for opening to world markets.

The lack of common interests within the coalition government was creating instability and becoming an embarrassment to Ecevit, whose policies were consistently obstructed by his deputy prime minister Erbakan and the NSP. With the Cyprus invasion in July 1974, however, the government and especially Ecevit personally, gained credibility and popularity. Ecevit decided that his popularity was now such that he could now safely hold early elections, dispense with Erbakan and form a stable RPP government. In September 1974 Ecevit resigned but he soon discovered that the rightwing political parties would not allow him to hold early elections from which they would surely emerge as losers.

The wrangle over the formation of the next government lasted for another six months during which Ecevit's chance slipped away. In March 1975, Demirel finally declared that he was ready to form the first Nationalist Front government by allying all rightwing parties, including the NSP, in one coalition. Dominant interests were less than happy with this solution to the immediate problem, since they had pushed for a RPP-JP coalition which would have had a very broad popular base, and was seen as the only feasible option for the maintenance of a stable Turkish state. Given the class bases of their parties, and the personal ambitions of Ecevit and Demirel respectively, this proposition was not a practical one since both had much to lose.

At this juncture, Demirel tried to satisfy the whole of the right by adopting a policy of uncompromising onslaught against the left and the liberals. With this drastic shift in policy, the seeds for the eventual dissolution of Turkish democracy were sown. The left was initially crushed after the military coup in 1971. Demirel now launched a pre-emptive strike against the left of the future by "overlooking" the violent activities of the "Grey Wolves" - the fascist activists of the NAP. By this time, the rightwing of the RPP had already defected to the ranks of the coalition of the JP, the NSP, the NAP and also the Democratic party and the Reliance party. Deals were made so openly on the floor of parliament in order to attract MPs away from the RPP ranks on to the Nationalist Front benches, that the event (neither the first nor the last time in Turkey) was dubbed the "MPs market". The fear of an early election which might return Ecevit to power, and the constant pressure to appeal to dominant interests, led the coalition steadily to the extreme right, thus strengthening the strategic position of the NAP and its leader Turkes in the coalition as well as in the country.

The state apparatuses were steadily infiltrated by the rank and file of the JP, the NAP and the NSP. Those who were opposed or unsympathetic to the Nationalist Front government were summarily dismissed, beaten up, harassed or openly killed. Demirel, whose political survival depended on the support of the extreme right, claimed that these deeds were being committed by "patriots" who were playing a supportive role to the security forces. He also said that the country was being ripped apart by the "Godless and Moscow-controlled" anarchists and, thus the integrity of the Turkish state was under threat. The acts of "patriots", therefore, were justified since they were aimed at the revival of Turkish nationalism and the protection of the Turkish state. These views were, naturally, echoed by Erbakan, Turkes and the other minor coalition partners. At this juncture, the NAP was pursuing a clearly fascist programme, and had taken control of the ministries of education and internal affairs, using force and intimidation when necessary. Turkes himself was personally responsible for the national intelligence organisation (MIT) where he had placed his own men in key positions and which he began to use as a source of personal power. [9]

The NAP played a significant role in the Nationalist Front government, in that while it was ostensibly contributing to the maintenance of a democratic system, it was actually engaged in the rapid dismantling of democratic processes and institutions. Demirel was so dependent on the NAP, however, that any policy difference that might have existed between the two parties became purely academic. The RPP, as the opposition party, was highly critical of these developments, and cultivated close links with DISK, the leftwing trade union federation, in order to create a base for an effective opposition to the growing trend of fascist practices. Ecevit was popularly seen as the last hope of Turkish democracy in stopping the rise of fascism.

As the Nationalist Front government increasingly condoned or promoted fascist practices, the dominant interests became increasingly concerned that stable government and economic growth could not be achieved by a ragbag of a coalition whose raison d'etre was violence. Monopoly/industrial and financial capital wanted an early election, and an end to the escalating violence and slaughter that was destabilising the nation. An early election was soon in prospect as, quite apart from its political excesses, the Nationalist Front government, with its confusion of policies, had mismanaged the economy so badly that foreign debts were mounting sharply, inflation had become rampant, foreign loans were scarce, and essential imports had reached an all time low. The country was, in short, on the verge of bankruptcy.

Elections were held on 5th June, 1977, and Ecevit's hope of achieving a simple majority in parliament was again disappointed. The RPP won 213 seats, which was 13 short of a majority. Ecevit was asked, nevertheless, to form a government, by the President of the Republic, and so the first minority government in Turkey's history came into being. The election also resulted in the demise of the small rightwing parties based on support from the petty-bourgeoisie. The bulk of the rightwing vote had shifted to JP and NAP, while NSP's vote was halved. This shift was not sufficient to form a strong rightwing government, but it was sufficient for the right to gain control of parliament. In addition to promoting fascist activities in the streets of Turkey, Demirel and Turkes effectively prevented Ecevit's minority government from functioning. When it came to a vote of confidence, the government was not able to obtain the required majority and was duly defeated. Demirel was then asked to form a new government.

The second Nationalist Front government came into existence as a coalition of the JP, the NAP and the NSP, but it did not last because the fascist practices of the NAP, especially the racist attacks against the Kurds and the Shi'as in the East and the Southeast, shocked some of the JP MPs sufficiently for them to leave the party to become Independent members. The government fell as a result of these defections, and Ecevit announced in December 1977 that he would form a government supported by Independents as well as by the remnants of two small, very conservative political parties. (These were the Democratic Party and the Reliance Party, which had taken part in the first Nationalist Front government (1975-77), but were almost wiped out subsequently at the elections). Thus, Ecevit's ambitions were at last realised, but the price of achieving them was too high. Ecevit promised "peace and unity" in the country and urgent solutions to Turkey's economic problems, but these promises proved empty. The fascist groups who now controlled key areas of government and bureaucracy stepped up their activities, leftwing groups reacted, and violence erupted all over the country.

Terrorism and random violence reached frightening heights, and the death toll increased daily. Although both left and rightwing groups were involved in violence, the police and other security forces largely ignored or protected rightwing terrorists while the left was mercilessly harassed. When the police intervened in a street battle, for example, in the great majority of the cases, the left and the liberal elements, or even passers by wounded by rightwing extremists, were arrested, imprisoned and tortured while fascists were allowed to escape. Where notable rightwing murderers were arrested, amazing prison escapes could be effected - as in the case of Mehmet Ali Agca, who later shot the Pope. The security forces had been extensively infiltrated by NAP and JP supporters during the previous Nationalist Front governments' days in power and they persistently disobeyed the orders of the RPP government. There was a concerted effort, which was obvious even to ordinary people, to undermine the legitimacy and effectiveness of the Ecevit government.

In December 1978 there was a major outbreak of violence, leaving many dead, in the city of Kahramanmaras, which promptly led to the declaration of martial law in thirteen provinces. [10] Ecevit managed to force the military commanders to operate within the framework of democratic principles, an action which did not exactly endear him to the military.

While terrorism was on the increase, the promised solutions to economic problems were still not forthcoming. In fact, there was precious little that Ecevit could do to alleviate Turkey's economic problems unless he was prepared to accept an IMF austerity package which would embark the Turkish economy on a transformation into an outward-looking system. Since Ecevit was not prepared to accept the package, the IMF was not willing to grant urgently needed loans. Without IMF approval, international banks were not willing to extend further credit facilities to Turkey, either. The economic situation was rapidly approaching a crisis point. The IMF wanted Turkey to devalue the Turkish Lira, to lower wages, to increase taxes, and to give priority to the export sector. All these measures were more than acceptable to dominant fractions of the Turkish bourgeoisie as long as loans were forthcoming and access to world markets secured.

Ecevit's last term in office was characterised by general frustration, but also by a critical lack of understanding of the needs of Turkish capitalism, resulting in a weak and disoriented government. As terrorism and economic hardship increased, and turned daily life for millions of people into a struggle for survival, Ecevit's popular support declined rapidly and steadily. When mid-term senate and national assembly elections were held in October 1979, support for Ecevit dropped sharply, while the JP and NAP increased their votes at the expense of the small rightwing parties. The 1979 election signalled a turning point in Turkish party politics in this respect since, for the first time, rightwing voters clearly opted for the major rightwing parties.

Ecevit resigned, and Demirel formed a minority government in November 1979. The NAP and NSP did not share directly in this government, but supported it in parliament, and retained their position in the state apparatuses which they had infiltrated earlier. Ecevit's resignation did not enable Demirel to cope with the continued escalation of violence, and the steadily worsening economic situation. The army was given a free hand in helping to maintain social and political stability. The IMF package, which Ecevit had resisted, was accepted in full, in January 1980, and monetarist policies were adopted. [11] Turgut Ozal, who had become the head of the State Planning Office and Demirel's right hand man, was given the task of overseeing the implementation of the 1980 IMF stabilisation programme. [12] It was clear, however, that the existing parliamentary system no longer provided a framework within which these crises could be controlled. On 12th September, 1980 the political transformation was completed by military intervention. After experiments covering a period of two decades, Turkey's parliamentary democracy finally came to an end.

THE IMF INTERVENTION OF 1979

The IMF intervention of 1979 was the final factor which propelled Turkey into the international arena of economic competition, and ensured the death of indigenous democracy. [13] As I have argued so far, Turkey was not ready to take on the challenge of operating immediately in international markets because of the problems created by its etatist and inward-looking structures. The contradictions of Turkish capitalism had reached a critical point where the opening of the economy and society, regardless its immediate and medium term costs, was the only feasible option available to dominant fractions of the Turkish ruling classes, i.e. monopoly/industrial and financial capital. The growing bottlenecks in the domestic economy, the steadily deteriorating international markets for Turkish products, the adverse effects of the 1973 rise in oil prices and the increasing unwillingness of centres of international capital to bail out Turkey, all made it imperative that Turkey's position in the world capitalist system be radically restructured.

The IMF intervention was not a sudden event, as the need for an externally imposed programme had been obvious to the dominant fractions of the ruling classes for some time. It had become increasingly clear from the early 1970s onwards that agricultural and commercial interests, and the growing working class were obstructing the opening of the economy, but it was also clear that overcoming these obstacles would entail fundamental social and political changes.

From the late 1960s and early 1970s, several influential centres of international capital began to put forward ideas on the international economy, which were directly opposed to the, so called, Keynesian principles which had been fundamental to the direction of the post-World War II international economic system. With the rise of the international oil crisis to the forefront of international politics, and the formation of OPEC as a new political force, these ideas became the new "received wisdom" of international economic and political strategies. The origin of these critical views can be traced back to the days of laissez faire economics of the 18th Century, but the new version of these old principles was marketed as the new economic strategy of "monetarism".

The IMF was one centre where monetarist economic and political strategies were well received. The world economy had previously been conceived as a world capitalist system based on the maintenance of individual national economies with corresponding nation-states (Brett, 1983). The new vision of the world was one in which existing economies would be rationalised and remodelled into a world economy; a single economic system with its various national economies serving this global economy as if they were its sectors, all operating to satisfy the demands generated within the world system rather than within the national economy.

This view of the world economy was bound to generate enormous opposition, and when the IMF stabilisation package was first suggested to the then Turkish government in 1979, Prime Minister Ecevit retorted with a fierce denounciation of this package and of the role of the IMF as the exploiter of the poor countries of the Third World. There was considerable justification in Ecevit's rejection of the IMF package on the grounds that it was totally unsuitable to Turkish needs and conditions. The conditions that Ecevit envisaged were those which had prevailed in Turkey since the establishment of etatism. The IMF package, however, was based on different assumptions and was aimed at total transformation of the Turkish economy within the world system, with no regard for the internal social and political repercussions within Turkey. Ecevit's government found itself under increasing pressure to accept this monetarist package as it failed to offer an alternative way out of the growing crisis.

The government was put in a very difficult political and economic situation. The legitimacy and the effectiveness of the parliamentary regime was tested daily in the streets of Turkey and found to be wanting. Parliamentary opposition by the JP, NAP, NSP and other minor political parties, not only brought the political system to a standstill, but also brought the Parliament into disrepute through the blatant pursuit of individual self-interest. The business community stopped investing as confidence plummeted, and instead hoarded available funds, while economic activity stagnated. International agencies increased the pressure on Ecevit to succumb to the IMF. Centres of international capital stopped trading with Turkey; the cheques issued by Turkish banks had become worthless. The international business community lost confidence in Turkey, Turkey slipped steadily in international credit ratings and found it virtually impossible to obtain credits to service its debts, let alone to find new sources of finance.

This combination of pressures on the government forced Ecevit to look elsewhere for support. He floated the idea that Turkey had long ignored its role as a natural supporter of Third World causes, and it was now time to assume its position in the Third World camp and to fight the growing threat of imperialism, especially from the United States. These views were a rationalisation for opposition to the IMF rather than a genuine programme, however, and had little general appeal. When Ecevit's government fell, and Demirel formed the next government, which had the support of the dominant fractions fo capital, but was, nevertheless, unable to meet the conditions demanded by the IMF. The inherent structural problems in the uneven development of Turkish capitalism still had to be resolved before the on the IMF's stabilisation programme could be successfully implemented.

The IMF was asking the Turkish government to undertake a series of policy decisions which were largely foreign to the nature of the Turkish social formation, but were very much in line with policy packages imposed on other countries. The IMF's aim was to integrate Turkey more firmly into the world capitalist system, and to place it in a position which would be conducive to the restructuring of the world economy along the lines indicated above. The suggested measures, therefore, were not geared to fit in with Turkey's existing economic and political structures, but instead were aimed at transforming them and bringing them into line with the pre-requisites of the world capitalist system. When we look at what these measures were, we can perhaps appreciate the situation more clearly.

The IMF proposals were:

a) to put greater emphasis on supporting the export sector.

b) to devalue the Turkish Lira. The Turkish lira had previously been kept at unrealisticly high levels in order to facilitate the relatively cheap import of manufactured goods in line with import-substitution policies.

c) to open the economy to international competition and to encourage foreign capital.

d) to introduce policies which would lead to the transformation of the agricultural sector.

e) to change the taxation system in order to encourage the further accumulation of capital.

f) to establish and maintain political stability.

As it is indicated in Chapter 1, there are pressures within the world economic system to impose uniformity on its parts, in order to preserve the continuous reproduction of capitalist relations on a world scale. It is, therefore, not surprising to find the imposition of strict and rather limited options on the constituent social formations of the world capitalist system. The strength and the weakness of the system lies in its ability or inability to achieve this degree of uniformity on a world scale. The most dangerous variable in this chess game, which is still being played today, is the political resistance, which is bound to be created within the different parts of the system against the concentration of capital into fewer hands, and against the increasing unevenness of the levels of absolute capitalist development. It is my contention that this resistance will take a military character and may ultimately signal the end of the world capitalist system. The IMF may be an important agency of world economic change, but it is very unlikely to prevent international collapse, if it is dominated by the interests of very few advanced and powerful capitalist countries to the exclusion of the growing and vocal opposition emerging in the Newly Industrialising Countries all over the world.

NOTES

1. See Ramazanoglu, 1985 and also, Hershlag, 1968; Walstedt, 1980; Hale, 1981 contains useful information about the Turkish Economy.

2. Turkish Industrialists and Businessmen's Association (TUSIAD), 1981, p.213.

3. See C. Ramazanoglu, 1985.

4. A detailed examination of the role of state economic enterprises in Turkey's economic development can be found in Walstedt, 1980.

5. A different view to the argument put forward here can be found in Berberoglu, 1982, where he argues on the basis of dependency school assumptions, that Turkey has not developed because of its neo-colonialist dependence of international, especially American, capital. He argues explicitly that imperialism has prevented the development of Turkish capitalism.

6. For a useful account of the 1971 coup, the nature of the military regime and the main socio-economic and political developments of this period, see Heinrich and Roth, 1973; Cem, 1977, and Ahmad, 1977.

7. An examination of this period and of the rise in political violence, can be found in Ahmad, 1973, pp.13-16.

8. For a brief examination of the gradual but persistent erosion of democratic and civil rights, which reached a peak after the 1980 military coup, see Ersan, 1981.

9. See Keyder, 1979, pp.35-41.

10. An analysis of this event within the context of Turkish capitalist development can be found in Birikim, 1979.

11. For examinations of the genesis of the present crisis of the Turkish economy, see Ebiri, 1980; Keyder, 1979; Balassa, 1981a and Balassa, 1981b.

12. For an account of the impact of the new stabilisation programme, see Senses, 1981.

13. A series of critical essays of the IMF policies and their impact on Turkey have been collected in Erdost, 1982 and also Dogan, 1980.

REFERENCES

Ahmad, Feroz (1973); "The Turkish guerrillas: Symptom of a Deeper Malaise", New Middle East, April

Ahmad, Feroz (1977); The Turkish Experiment in Democracy 1950-1975, Hurst, London

Ahmad, Feroz (1981); "Military Intervention and the Crisis in Turkey", MERIP Reports, No.93, pp.5-24; also Chapter 7 of this book

Balassa, Bela (1981a); "Policies for Stable Economic Growth in Turkey", in B. Balassa, The Newly Industrialising Countries in the World Economy, Pergamon, New York

Balassa, Bela (1981b); The Policy Experiences of Newly Industrialising Economies After 1973 and the Case of Turkey, paper presented in Istanbul, 1981

Berberoglu, Berch (1982); Turkey in Crisis, Zed Press, London

Birikim (1979); Maras'tan Sonra (After Maras), Birikim, Istanbul

Brett, Edward (1983); International Money and Capitalist Crisis, Heinemann, London

Bulutoglu, Kenan (1970); Turkiye'de Yabance Sermaye (Foreign Capital in Turkey), Gercek Yayinevi, Istanbul

Bulutoglu, Kenan (1980); Bunalim ve Cikis (Crisis and Escape), Tekin Yayinevi, Istanbul

Cem, Ismail (1973); 12 Mart (12th March), Cem Yayinevi, Istanbul

Dogan, Yalcin (1980); IMF Kiskacinda Turkiye, 1946-1980 (Turkey in the Stranglehold of the IMF, 1946-1980), Toplum, Ankara

Ebiri, Kutlay (1980); "Turkish Apertura", METU Studies in Development, Vol.7, Nos.3-4, pp.209-52; also Chapter 4 of this book

Erdost, Cevdet (ed.) (1982); IMF, Istikrar Politikalari ve Turkiye (IMF Stabilisation Policies and Turkey), Savas Yayinlari, Ankara

Ersan, Tosun (1981); "Turkey's Battered Democracy", INDEX on Censorship, Vol.11, No.126, pp.11-15

Goymen, Korel and Tuzun, Gurel (1976); "Foreign Capital in Turkey: An analysis of Capital Imported Under the Encouragement Law, 6224", Studies in Development, No.11, pp.55-79

Hale, William (1981); The Political and Economic Development of Modern Turkey, Croom Helm, London

Harris, George (1970); "The Causes of the 1960 Revolution in Turkey", Middle East Journal, Autumn

Heinrich, Brigitte and Roth, Jurgen (1973); Partner Turkei oder Foltern fur die Freiheit des Westens?, Rowohlt, Hamburg

Hershlag, Z.Y (1968); Turkey: The Challenge of Growth, Brill, Leiden

Keyder, Caglar (1979); "The Political Economy of Turkish Democracy", New Left Review, No.115, pp.3-44

Landau, Jacob (1981); Pan-Turkism in Turkey: A Study in Irredentism, Hurst, London

Ramazanoglu, Caroline (1985); "Labour Migration in the Development of Turkish Capitalism", Chapter 6 of this book

Ramazanoglu, Huseyin (1985); "A Political Analysis of the Emergence of Turkish Capitalism, 1839-1950", Chapter 2 of this book

Samim, Ahmet (1981); "The Tragedy of the Turkish Left", New Left Review, No.126, pp.60-85

Senses, Fikret (1981); "Short-term Stabilisation Policies in a Developing Economy: The Turkish Experience in 1980 in Long-term Perspective", METU Studies in Development, Vol.8, Nos.1-2, pp.409-451; also Chapter 5 of this book

Turkes, Alparslan (1977); Milli Doktrin, Dokuz Isik (National Doctrine, Nine Lights), Kutlug Yayinlari, Ankara

TUSIAD (1980); The Turkish Economy, TUSIAD, Istanbul

TUSIAD (1981); The Turkish Economy, TUSIAD, Istanbul

Uras, T. Gungor (1979); Turkiye'de Yabanci Sermaye Yatirimlari, (Foreign Capital Investments in Turkey), Iktisadi Yayinlar, Istanbul

Walstedt, Bertil (1980); State Manufacturing Enterprises in a Mixed Economy: The Turkish Case, World Bank/Johns Hopkins University Press,

Baltimore

Weiker, Walter (1963); The Turkish Revolution 1960-1961, The Brookings Institution, Washington D.C.

World Bank (1975); Turkey: Prospects and Problems of an Expanding Economy, Washington D.C.

4 Turkish apertura*

KUTLAY EBIRI

1. INTRODUCTION

As the crisis of the Turkish economy was reaching unbearable proportions towards the end of 1970s, "official" Turkish development strategy was being subjected to widespread criticism, and a rather belated debate on the external framework of development was coming onto the scene. The by-elections in October 1979 gave way to a new government which formulated a stabilisation programme, the so-called "25 January decrees". The stabilisation programme consisted of four elements: (i) a more realistic exchange rate, (ii) a tighter monetary policy coupled with higher and more flexible interest rates, (iii) some measures for export promotion, mainly reduction in "red tape", (iv) a small drift from etatism, i.e. the traditional strategy of direct economic activity by state enterprises as well as comprehensive regulation by the government. However, after the programme was declared, the debate lost much of its economic content and was turned into a rather uninformed version of party-political struggle. While the "25 January programme" was in action, mounting political violence led to a military takeover in September 1980. The government appointed by the military leaders had Turgut Ozal, the "engineer" of the stabilisation programme, as the Deputy Prime Minister. The debate on the development strategy in its politically motivated form seems to have died down. However, the conditions which created the debate on the strategy of development in the first place are still there in full force; and although the party-political characteristics of the arguments have been weakened, cliches and dated catchwords still dominate the scene.

This article attempts to provide a systematic introduction to the

---

* This is an edited version of the article first published in METU Studies in Development, Vol.7, No.3-4, 1980.

Apertura is the shortest expression of the process of transition from inward-looking to outward-looking development strategy. Spanish.

main issues of the debate. It aims to set the scene for the analysis of the arguments for and against "apertura", including the "para-economic" aspects of the controversy.

It consists of seven sections. In Section 2, we present a conceptual treatment of the external dimensions of development strategies. Section 3 attempts to classify the explanations offered for the recent economic crisis in Turkey. Turkish traditional development strategy of import-substitution is described briefly in Section 4, and reference is made to the attempts of "reformulation". Section 5 provides an overview of "apertura", and refers to the conditions leading to a more outward-looking framework; while Section 6 attempts to identify the "sources" of anti-apertura arguments.

Three different arguments are raised against "apertura": (i) "apertura" makes the country "more dependent on Western capitalism", (ii) an outward-looking strategy would obstruct "real industrialisation", and (iii) it would worsen "income distribution" and therefore require "political repression". As the "dependence" argument is based upon a mixture of political, ideological and economic considerations and provides "inputs" for other anti-apertura arguments, it is considered more appropriate to treat it in Section 7.

Two restrictions about the subject matter of this article must be noted at the outset. First, the scope is limited, to the extent possible, to the external framework of development strategy. Thus the equally important controversies on the virtues of state intervention, the causes and consequences of inflation, and the methods and institutional framework of development planning are left out. Complete isolation is impossible however, since most arguments studied in this article are originally put forward incorporating the aforementioned issues. Secondly, this study deals mainly with post-1977 debates. Many topics discussed here have been the subject of earlier discussions on development strategy, especially during the 1930s and in the late 1960s, in connection with Turkey's associate membership to the EEC. Yet the basic problem of Turkish industrialisation emerged clearly only after 1977; all sources of external financing were either exhausted or rendered insignificant by the proportions of the trade gap reached by 1977, with attempts to get more from these "usual" sources largely frustrated. Dimensions of the problem were far from the reach of marginal adjustment in export incentives and/or import restrictions, bringing a major overhaul and a very radical transformation of the development strategy to the agenda. Some, the author included (Ebiri, 1979b), argued that import-substituting-industrialisation was at a deadlock, it had already sung its swansong during 1974-77 and all that was needed was "euthanasia" to open the way for an outward-looking path of development; while others attributed the problems face not to the strategy of import-substitution but to a series of external factors and to the "mismanagement" and "mistakes" in the application of that strategy. The proponents of the latter argument were opposed to "apertura" with such adamance that some went as far as proposing an equally radical strategy of self-reliance. Post-1977 debate on development strategy has thus seen both strategies put forward in their radical forms rather than as minor amendments to the traditional strategy.

2. EXTERNAL STRATEGY

External strategy has three dimensions in general: (i) trade (ii) finance, and (iii) technology. Trade dimension is essentially related to the question (and the decision) whether the production structure of the country is to be designed that the country will take part in the international division of labour - either immediately or over a certain period, or whether production (in terms of quantity, quality and cost) will be directed mainly to the domestic market. External financial strategy is concerned with the question of the role of international capital (private and public) in the country's development programme. Thus a country may choose to rely mainly on the domestic financial resources either by forced savings or by reducing the growth rate to a level where it can be financed by voluntary savings or may decide to borrow to bridge the gap between actual and required savings. When the first course is adopted, i.e. when international capital is not used, foreign exchange component of investments (the part which is not substitutable by domestic production) must, of course, be financed by export revenue.

The last dimension - technology - has more far reaching longer term implications: and outward-looking technological strategy consists of keeping a sharp watch-out for technological developments worldwide with a view to transfer, adopt, assimilate and finally to contribute to modern technology, though not necessarily in all lines of production. An inward-looking technological strategy on the other hand, while not as demanding, may be very "tiresome" as a large part of the country's resources under this strategy must be devoted to rediscovery of the already beaten tracks of world technology. [1]

It is easy to see that a country may be outward-looking in relation to one of these dimensions while inward-looking in others and *vice versa*. Soviet industrialisation experience, for instance, is an example of an inward-looking strategy in terms of trade and financial resources but outward-looking in terms of technology. [2] Japanese experience is another example of a mixture: outward-looking in terms of trade - especially export trade - and technology, but inward-looking in financial resources. [3]

Of prime importance for the present study is the case where an inward-looking production structure and technological strategy are combined with an outward-looking financial strategy. Development strategy of Turkey, like those of many other developing countries does, or has to, rely on external financial support extensively in order to maintain the production of and investment into the inward-looking industries which cannot earn the necessary foreign exchange. I refer to the aforementioned combination as the "import-substituting-industrialisation" strategy.

On the other hand, if one is to remain faithful to the original aim of import-substitution strategy (or to the literary meaning of the term) it may be difficult to call a strategy "import-substituting" when it causes the foreign exchange gap to widen, instead of bringing about a more self-sufficient trade structure. Indeed, those who have attempted to measure the "degree" of import-substitution have used the change in the ratio of imports, direct or indirect, to total supply as the basic indicator. [4]

Yet when defined and measured by import/supply ratio, import-substitution may represent just another fame for industrialisation in general. (Or even "agriculturalisation" when a country's imports initially consist primarily of agricultural products.) In other words, import-substituting-industrialisation loses the chance of being a policy alternative, but becomes an all-embracing concept including all kinds of structural change. Let us see why.

To start in a given country to produce something which is not previously produced there, is "import-substitution", for it is almost certain that it was being imported previously in small or large quantities. When measured in terms of direct import/supply ratio, import-substitution will be empirically observed. When direct imports (i.e. imports of inputs required by the new line of production) are taken into account, the import/supply ratio becomes slightly more indicative of a policy choice: "resource based" industrialisation (including both agricultural and mineral processing) may not raise the imported input requirements [5] and thus may represent a non-import-substituting option. But even by including the indirect imports, one cannot avoid recording the most export-oriented activity of all, namely, multinational subcontracting arrangements as "import-substitution"! On the other hand, the substitution measure, taking into account both direct and indirect imports, will not allow us to classify a long-term, deliberate inward-looking industrialisation strategy as import-substituting-industrialisation, so long as it remains at the stage where intermediate input imports rise faster than the domestic output of the final products. It may be argued, of course, that, from the point of view of the economy as a whole, such considerations matter little, for if the ratio of aggregate imports to national income rises over time this must be taken to represent a stronger relationship with the external world, and certainly not as import-substitution. The latter will be identified only with a declining import ratio. The implicit assumption is, of course, that a rise in import ratio cannot be sustained for a long period without being accompanied by a rising export ratio. Yet the reverse being the common case for a large group of developing countries following an inward-looking path of industrialisation, there is little justification for this assumption. We must conclude, therefore, that the accepted measure of import-substitution does not reflect the industrialisation policy adopted, i.e. it is not useful in distinguishing between export-oriented and import-substituting strategies. Common practice is to distinguish between these two strategies without strict adherence to the changes in import ratios. Thus strategies favouring production for domestic use - and discriminating against export activities - through various measures (extending from exchange rate policy to ad hoc factor price distortions) and resorting to external borrowing to bridge the resource gap are commonly called "import-substituting-industrialisation". [6]

Export-oriented strategies clearly are not their diametrical opposites, for although the promotion of exports is chosen as the central idea, they make use of foreign private and public capital. Further, many characteristics of import-substituting-industrialisation survive under export-promoting-strategies, when the latter follows the former in a chronological sense. In fact, this has led many authors to the view that import-substitution and export-promotion are complementary and that the "dichotomy" is false. [7]

Going back to the general framework of our analysis, viz. inward and outward-looking strategies, we can see easily that an export-promoting industrialisation strategy need not close its doors to foreign capital even when it succeeds to bridge the trade gap completely. Rising exports generally create better conditions to borrow and hence to further expand the capacity to import. Furthermore, contrary to what was widely believed a decade ago, export-promotion strategies do not have to rely on "backward" techniques. In fact, many cases, export-oriented economies have proved to be better followers of worldwide technological progress.

3. EXPLANATIONS OF THE RECENT ECONOMIC CRISIS

In order to be able to accommodate the arguments for and against Turkish "apertura", one has to go back to the explanations offered for the recent crisis of the Turkish economy. In this section, I shall try to summarise the main theses on crisis to the extent they are related to concrete policy proposals hence forming particular points of view as regards "apertura".

Although not wholly reflecting the division between those "for" and "against" a more outward-looking economy, "short-term" and "long-term" explanations can be easily identified and it is illuminating to see the correspondence between the short-run explanations and the arguments against "apertura". The "standard" short-run cause is "rising oil prices": Turkish growth was halted, as the price of oil quadrupled and together with the induced rises in other industrial prices, import bill rose to a very high level which could not be financed by "normal" sources of foreign exchange. Government resorted to short-term credit and managed to bridge the current account deficit without curbing the demand. But the oil price continued to rise - with it, all other prices. Because Turkey could not keep its dues, she was not able to obtain additional credit, whatever the terms. By the end of 1977, therefore, Turkish industry went into severe difficulties, shortages developed in many intermediate and consumer goods and the rate of inflation soared.

This "standard" and often "official" (SPO, 1979) explanations is sometimes supported by other short-term "factors" like "economic mismanagement by consecutive short-lived governments" [8], "the rise in terrorist activity", frequently interpreted as an extension of a longer term factor, viz. "an international grand conspiracy". The latter seems to be a very attractive "explanation" and it has many followers. Therefore, I shall dwell on "conspiracy" factor at some length among longer term explanations. A less developed and shorter term version of "conspiracy" attributes Turkey's problems to the chain of events following the Turkish military intervention in Cyprus. Cooler diplomatic relations with the Western world and American military embargo are forwarded as the factors contributing to the prolongation of the crisis. [9]

Three different long-term explanations of the present crisis of the Turkish economy can be identified: "capitalist dependence", "international conspiracy", "import-substituting-industrialisation".

"Capitalist dependence" argument is a simple extension of the Marxian theories of imperialism. [10] The present crisis, as well as the structural, or perhaps "permanent" underdevelopment, are interpreted as direct outcomes of the links between international capitalism and the Turkish economy. Although widely divergent in their assessment of the effects of these relationships on the Turkish economy, almost all versions of Turkish "dependencia" conclude that the present difficulties emanate from the crisis the world capitalism is experiencing. [11] Since crisis is assumed to be a phenomenon which accompanies, or is embodied by capitalism, Turkey cannot expect to solve her own problems without severing her links with the world capitalist system (Altintas, 1978; Bortucene, 1979:23-25, 37-38; Erdogdu, 1978:43-4, 56-57 and 1979:19-29; Galip, 1979; Kafaoglu, 1979; Ozkol, 1969 and 1970; Sonmez, 1980).

"International conspiracy" is often forwarded as a supporting element for "dependencia" explanation, but it certainly has its adherents in its own right. The commonest version is this: Turkey was rapidly growing and industrialising thanks to its import-substituting policy under <u>etatist</u> guidance. This has disturbed the industrial world and they have acted, through the IMF "to draw the line" (Boratav, 1979). [12] The IMF, "in fact" has been trying to hinder Turkish development efforts since the beginning of the 1950s: devaluations are forced upon Turkish governments in order to serve the interests of the Western countries (Dogan, 1980: passim; Kafaoglu, 1979); the World Bank tries to reduce growth rate of the Turkish economy (Ozgur, 1976; Turkcan, 1978).

Both "capitalist dependence" and "international conspiracy" explanations are widely accepted - they both take the responsibility of the crisis away from those who have strong vested interests in the policies followed and from economic decision makers - and both explanations are used, although somewhat naively, to form the arguments against "apertura".

The third long-term explanation of the recent crisis attributes the present difficulties to the "import-substituting-industrialisation". However, two different versions of the latter must be distinguished, for they lead to opposite policy proposals in terms of external strategy. First version takes the view that import-substitution started from "the wrong end", i.e. from consumer goods, contrary to the "principles" adopted in Five-Year Plans. It was also executed without a conscious programme, that is, it got out of hand, or (alternatively) was applied "excessively", resources were too thinly spread, scales were too small, etc. In other words, had Turkish import-substitution followed the planned path, or been more moderate, the crisis would not have come, even the world economic crisis would not have affected the Turkish economy. This version is sometimes forwarded in support of "capitalist dependence" and/or "international conspiracy" explanations, too (Akgul, 1976; Boratav, 1979; Demirer, 1977; Erdem, 1980; Kandiller, 1976; Sonmez, 1980; Korum, 1976; Kucuk, 1980:482-3; Tuzun, 1979; Yalin, 1980).

Second version maintains that Turkish import-substitution has developed in the same manner that it did in other developing countries, i.e. it rejects the view that import-substitution started from the "wrong end" or that "it got out of hand". Import-substitution strategy consists, by definition, of the domestic production of consumer goods at first, then of the half-hearted attempts to produce intermediate and capital goods, with the technology and scale being determined by the protective structure of import-substituting industrialisation and by the specific distortions created in relative factor prices. Turkish import-substitution had to start from "there", too, given the socio-political framework of the country. Further, the policy to push import-substitution "backwards", i.e. to intermediate and capital goods, to the extent that it was successful, deepened the crisis (instead of strengthening the structure of industry, as implied by the first version).

Turkish import-substitution enjoyed two special advantages: first, the traditional central control of the economy created an environment where most distortions escaped public attention or were accepted as "facts of life", and secondly, large sums of migrant workers' remittances enable the Turkish economy to survive a large and widening trade gap. Both phenomena postponed the crisis of import-substitution until 1974, when oil and other import price rises extended the dimensions of the problems created by the inward-looking trade strategy. In fact, the latter had already destroyed the "shock absorbers" and the Turkish economy responded to the oil crisis by adopting policies diametrically opposite to those appropriate under such conditions (Dervis and Robinson, 1978:38-71; Hatipoglu, 1978; Divitcioglu, 1980; Ebiri, 1979a:3-7 and 1979b; Kurdas, 1979; Okyar, 1976; World Bank, 1979:16-20).

All explanations outlined above lead, explicitly or implicitly, to anti-crisis policy proposals. Some of these are essentially political in character, some are short-term remedies and some long-term proposals imply far-reaching structural changes. "Apertura" proper belongs to the latter. The present programme put into action by the beginning of 1980 has not yet dismantled the basic mechanism of the import-substitution strategy, and it is not certain whether it aims at real outward-looking industrialisation. I assume in this paper, however, that the arguments for and against "apertura" in Turkey are put forward with a real and radical change in mind.

4. TRADITIONAL STRATEGY OF DEVELOPMENT IN TURKEY

Turkish development strategy in the post-war period is based upon the production of import-substitutes without reference to country's comparative advantage (of any sort). Thanks to the excessive external protection and to the "featherbedding" policy, the newly established import-substituting industrial (public and private) firms did not have to take into consideration the exploitation of the economies, the problem of optimality of the allocation of resources, or to worry about X-efficiency.

"Genesis" of Turkish import-substitution is sometimes taken back to the early 1930s. It is often argued that "the great crash" of 1929 which disrupted international trade and the termination of "Lausanne restrictions" [13] created the conditions for inward-looking industrialisation. It is true that the new 1929 tariff rates were considerably higher than the 1916 rates (Kurmus, 1978); however, the effective rates were similar to those prevailing in other European countries (Tekeli and Ilkin, 1977:70). More important is the fact that central allocations of resources, through the <u>etatist</u> framework, favoured resource based industries and agricultural processing (Cavdar, et. al., 1973:155-56; Malkoc et.al. 1973: passim; Rozaliev, 1978: 157-61). On the other hand, post-war substitution has clearly shifted the balance towards durable consumer goods industries which had heavy requirements of imported intermediates. Therefore, in terms of the dichotomy analysed here (i.e. import-substituting and export-promoting strategies) it is only the post-war strategy which is "worth" the name "import-substitution" (Boratav, 1979 and Gulalp, 1980). Pre-war industrialisation substituted the "basic needs" imports and did not have the other characteristics of typical "import-substituting-industrialisation" (see below, Section 7).

Import-substituting-industrialisation has formed the basic philosophy of development and was almost a "dogma" in the post-war developmental thinking in Turkey. All organised social groups accepted it as "the only way" of development. It was expected that this specific form of industrialisation would "free the country from the vagaries of world economy", would "close the structural foreign exchange gap", would "transform the predominantly agricultural labour force into an industrial proletariat", until it was realised in the past half-decade that these were pipe-dreams. The foreign exchange gap became ubiquitous, making the economy more vulnerable to the external "vagaries", and the industrial labour demand remained weak thereby causing a disproportionate rise in the numbers unemployed and underemployed.

The fact that the gross domestic product in Turkey grew rapidly over the post-war period is often used to show the "superiority" of the traditional import-substitution strategy in Turkey. [14] Yet to attribute this performance to inward-looking industrialisation, one needs to know whether the latter was able to create the resources to sustain the process of growth. The rate of growth of the domestic product over the period of 1965-79 is 6.2 per cent. During the same period, Turkey sold abroad some $17.5 billion worth of goods and imported $39 billion, leaving a trade gap of $21.6 billion. Thanks to the Turkish workers in Europe, who were there certainly not because of the success of the import-substitution but partly as a result of its failure to keep its pledge on employment, almost half of this trade gap ($10.7 billion) is paid without resorting to foreign borrowing. The rest ($10.9 billion), however, had to be met by foreign resources and although for some time it was possible to find loans on concessional terms, import-substituting extravaganza gradually became rather costly. Workers' remittances and foreign loans had to pay for the annual gaps created by the import-substituting strategy. A comparison of the first and sixth columns of (Table 1) indicates that the proportion of commodity resource balance (i.e. commodity imports-commodity exports) to gross domestic product is 5.2 per cent on average (unweighted) over the period, and that this accounts for as much as 84 per cent of the average annual growth rate of

TABLE 1

Growth, Foreign Trade and Resources, 1965-1979

Year	GDP Growth %	X	M	X - M	R	X - M GDP %	C
1965		502	645	- 143	80	1.9	9.00
1966	11.8	541	815	- 274	130	3.0	9.00
	3.8						
1967	7.5	554	836	- 282	123	2.8	9.00
1968	5.7	573	882	- 309	132	2.8	9.00
1969	5.3	622	993	- 371	181	3.0	9.00
1970	9.0	588	948	- 360	273	4.1	14.92
1971	6.0	677	1171	- 494	471	4.2	14.15
1972	4.4	885	1563	- 678	740	4.6	14.15
1973	8.5	1317	2086	- 769	1183	4.1	14.15
1974	8.8	1532	3777	- 2245	1426	8.5	13.99
1975	8.7	1401	4739	- 3338	1312	8.8	15.15
1976	5.7	1960	5129	- 3169	983	8.7	16.05
1977	3.7	1753	5796	- 4043	982	5.7	17.91
1978		2288	4599	- 2311	983	5.0	24.28
	.9						
1979		2261	5069	- 2808	1694	6.1	41.64

Symbols: X= Export of Goods, FOB, $ million
M= Imports of Goods, CIF, (Both X and M figures for 1965-69 period from Kayra's, 1972, adjusted data. He observes also that import prices are understated, p.7.
R= Workers' Remittances, $ million
C= Official value of one $ in Turkish currency (1970-79 values are weighted averages)

Sources: United Nations (1978:752), World Bank (1979), Central Bank (1980), State Institute of Statistics (1978), Kayra (1972:103/A)

the economy. Without, of course, a statistical relationship being suggested, one is nevertheless tempted to interpret these figures as most of the economic growth during this period being brought about by the funds not generated by the domestic economy. [15] This interpretation is strengthened when the overvaluation of the exchange rate and the illegal imports are taken into account.

Even before the oil price rise in 1974, traditional development strategy of Turkey required an import growth (in value terms) close to 15 per cent a year on average. On the other hand, insistence on a constant and overvalued exchange rate and the general bias against exports up to 1970 kept the growth of exports at a very low rate. Export revenue more than doubled, however, between 1970-74, following the devaluation of the lira, two bumper harvests in agriculture, and the fall in the value of the US dollar. Although there was only a slight increase in the share of manufacturing in total exports during this period, signs of outward-orientation were shown only by textiles and clothing industries. Other manufacturing exports have had a very hectic behaviour. Some half of the new products exported every year, albeit in small quantities, were not exported again, and some items, which were traditionally large foreign exchange earners, disappeared altogether from list (Kilickaya and Ibrahimhakkioglu, 1978; Togay, 1974). However, the expansion of Turkish exports after the devaluation and the rise in workers' remittances led some Turkish economists to believe that the foreign exchange problem was finally solved and that the economy would not be feeling the pressure of the "trade gap" for a long time to come (Ipekci and Ustunel, 1973). Others, however, were more cautious and pointed to the unique chance of the Turkish economy in 1973 to adopt a more liberal, more egalitarian and outward-looking strategy while foreign exchange reserves were rapidly accumulating (Dervis, 1974; Hatipoglu, 1973).

On the other hand, Turkish planners were quite explicit in their preference for import-substitution over export-promotion, during the preparation of the Third Five-Year Plan 1973-77 (Turel, 1972:88-91). The problem was partly that they regarded import-substitution as identical with industrialisation. Some thought import-substitution not as a specific strategy but as the reduction in the import/supply ratios, and therefore allowed themselves to speculate on "import-substitution plus export-promotion" programme, failing to realise that the actual strategy they were designing was heavily discriminating against exports (Cetin, 1972:39, 53). Further, the allocation of the global figure of "required substitution" among industries (Varlier, 1972) could only be called rudimentary (if not random). [16]

The Third Plan was based on a "new" strategic formulation which was claimed to be different from the "older" one on two counts:

(i) "New strategy" (NS) aimed at exploiting country's "development potential" to the full, while the "old strategy" (OS) tried to achieve the 7 per cent growth target by clearing the bottlenecks, reducing capital underutilisation and unemployment, and by diversifying exports.

(ii) NS aimed to reach the target level of income with "a certain industrial structure", claiming that this principle is adopted for the first time in Turkish planning. Since it is used as the basis of one of the arguments against "apertura", it is worthwhile to quote the NS definition of this "certain industrial structure":

"...It is aimed to reach an industrial structure where the shares of intermediate goods industries (which have strong backward and forward linkages) and of capital goods industries (which create employment, promote...technical change, contribute greatly to the national product, reduce dependence on foreign economies but raise Turkey to the position of a strong trade partner) are increased, with great care also being place upon the requirements of national security."
(SPO, 1973, p.984. My translation and emphasis, K.E.)

Although it is a matter of speculative interest whether the NS would indeed provide for a better exploitation of the "development potential" had the oil crisis not developed, there is nothing in the Third Plan to indicate a shift in the policies to utilise the economic resources more effectively. In fact, the Third Plan stipulated a lower marginal savings ratio, allocated more resources to those sectors which used capital less productively, and paid less attention to the problem of including a higher proportion of the working age population into productive employment (SPO/DSP, 1977).

On the other hand, the strange "priority" principle, which in effect, declared all manufacturing sectors except food and textiles as "priority sectors", plus others like energy, housing, etc. which are given priority separately in their own sections, did not mean very much in the actual construction of the consistency model. The latter allocated resources among sectors in order to satisfy the demand vectors projected on "technical" grounds, not incorporating such "priorities" as quoted above. Thus, in spite of the formulation of a "new" strategy, the share of planned capital goods investments in total fixed capital formation is 1.5 percentage points higher than in the Third Plan (than the corresponding ratio in the Second Plan)! [17] More interestingly, the 64-sector consistency model prepared for the Fourth Plan (which is supposed to represent the second stage of the "new" strategy) reduced that ratio by two percentage points. [18]

It is not difficult to see that behind the need to formulate a "new" strategy lie the problems created by the import-substituting-industrialisation. Yet the NS is designed with the traditional "autarkic" tendencies and failed to direct attention and organised effort to the basic issues, viz. the lagging exports, inefficiency in the allocation and use of economic resources, very thinly spread investment, etc...

What the planners and many economists failed to see were the links between the strategy of import-substitution and the pressing problems which presented themselves independently. Thus low agricultural productivity, energy bottleneck, inadequate infrastructure, distorted and energy-wasting transportation system, increasing dependence on foreign loans, rising unemployment, income inequality and inefficient public enterprises alongside a mediocre bureaucratic apparatus [19] were either considered to be "evils on their own right" or blamed on "capitalism" as a whole. Glorification of industry and the exaggeration of its advantages vis-a-vis agriculture plus the identification of import-substitution with industrialisation have obscured such links and raised import-substitution "above suspicion".

The standard argument in defence of the traditional inward-looking industrialisation strategy is that one must assess the costs and benefits of such programmes in the long-run. The expectation is that today's "infants" will grow and solve all problems. Yet, quite apart from the dynamic efficiency problem, the question is whether inward-looking strategies, can provide self-sustaining growth without requiring large sums of foreign loans and technology, that is, external support.

5. AN OVERVIEW OF "APERTURA"

In terms of the "trade" dimension of the external strategy, "apertura" consists of transforming the present structure of industrial production, gradually or rapidly, into one that is competitive in the world market. Whether this is to be done under increasing government intervention or by relying more in market signals is not the heart of the matter. Although, logically, extensive government control does not seem "to get on very well" with outward-looking strategies (Keesing, 1967) there are many examples of "closely watched" and government sponsored export programmes (Balassa, 1977; Berry and Thoumi, 1977; Wolf, et.al., 1977; Westphal and Kim, 1977).

A radical reorganisation of domestic production and technology, raising the production of export revenue to gross domestic product while "forcing" the domestic producers to allocate and employ the resources available in the most efficient combinations and tempo inevitably involves costs. "Gradualist" and "critical minimum effort", Leibenstein (1957:94-110), strategies may differ in terms of costs and is not possible to assert *a priori* that the former will always be less painful.

Although rarely brought into the discussion, "apertura" has a technological dimension, too. The country has to concentrate its scientific and technological capacity to outward-looking sectors to transfer and adapt the most advanced process and product technologies with a view to build upon the borrowed technology and know-how in order to support the export offensive. This is not necessarily in conflict with the pattern of specialisation (and the implied choice of techniques in terms of factor proportions) for it is well-known fact that capital-intensity and other technological characteristics of commodities do not have a one-to-one relationship (Hufbauer, 1970). Much depends on the process of adaptive innovation, product mix, quality and design considerations and the degree of interdependence between main and ancillary processes (Frankena, 1974).

Because import-substituting-industrialisation does not reduce (and in many cases increases) the need for foreign borrowing, the change in the financial dimension of the external strategy in "apertura" is relatively less important. The difference lies in the fact that a country with rising export revenue faces much more favourable conditions in the international capital market than those facing an import-substituting country which has to grow under the extreme pressure of large trade deficits. On the other hand, import-substituting countries proved to be equally attractive to foreign direct investments so long as they offered large and protected markets and so long as they were able "to keep their heads above the water".

Further comparisons between inward and outward-looking strategies can be made in terms of X-efficiency, economies of scale, capacity utilisation, job creation, income inequality, political structure (see Keesing, 1979, for an excellent treatment of these issues).

Since a large part of the controversy over "apertura" seems to stem from the confusion in the definition of the relevant alternatives, it may be appropriate to emphasise that the choice is not between a smoothly functioning import-substituting-industrialisation and an export-promoting strategy. As we have already seen, import-substitution is generally accepted to have encountered "not-so-very-short-term" difficulties and the arguments for "further import-substitution", for example, invariably imply very radical, if not politically impossible, policy changes. Further, what is at issue is not whether it would have been much better (or worse) had Turkey launched an export-oriented strategy in the post-war period. This could be a question of historical curiosity and speculation, but not a relevant one in a policy-oriented debate in 1980. The long process of import-substitution experienced created new conditions (favourable or unfavourable) and placed new constraints on policy options. This means, for example, that a wholly agriculture based "apertura" cannot be in the agenda of the discussion today. Building up of a large industrial capacity, albeit very uncompetitive by world standards, and a sizeable industrial labour force allow for (or perhaps dictate) an important part of future exports to come from the manufacturing sector.

Yet the present attempts at "apertura" in Turkey does not seem to be the result of the full realisation of the costs attached to import-substituting-industrialisation and hence latter's static and dynamic inefficiency, but the outcome of the deadlock created by the present crisis. In other words, "apertura" is attempted not because it is preferred to the traditional strategy, but because the latter is not feasible any more.

In fact, it would be naive to think that the politically influential beneficiaries of inward-looking and protectionist regimes would give up their privileges provided by the traditional strategy and "prefer" an outward-looking and more liberal external framework. "Apertura" is forced upon these groups by the crisis of import-substitution, and they regarded and supported it as a means of increasing total foreign exchange earnings of the country, without themselves being forced to export, and as a move towards establishing closer links with foreign capital (not necessarily for export-oriented ventures). [20]

6. SOURCES OF ANTI-APERTURA ARGUMENTS

Opposition to any change is expected to come from those who have vested interests in the status quo. Yet in a complex society it is extremely difficult to identify the links between purely economic interests and ideas to assign views to specific groups. Therefore, our treatment here of anti-apertura arguments is based on "issues" rather than on interests. On the other hand, in order to analyse the arguments we need to identify the sources from where these arguments stem. Thus to understand the anti-apertura argument centred around "dependency" theory, we have to a be aware of the idea of "nationalism" in its most basic form. Likewise, "export pessimism", which currently commands almost an axiomatic prestige, feeds the argument that "apertura obstructs real industrialisation" as well as providing some support to "dependency" argument, and so on.

Before going into the detailed analysis of anti-apertura arguments, therefore, these sources must be defined. The latter are described below in their value-laden forms, in order to reveal the lines of reasoning clearly. Of course, nobody defends these views in their crude and axiomatic forms, therefore, no references are made to authors in the following presentation. The following classification is not meant to represent mutually exclusive points of view, the reverse in often the case; arguments against "apertura" are usually made up of a mixture.

"Nationalism": Self-reliance (and even autarky) is preferable (ideal). International trade is harmful. Trade and growth are negatively correlated. Trade dependence is bad, just as any other dependence. Imports must be limited to the essentials, which are impossible to be produced within the country, and exports must be so much as to meet the import bill. A liberalised economy will cause the present industry go bankrupt. There exist in this world many countries which developed through inward-looking policies; what is more, Turkish import-substitution was very successful when it was under strict state control between 1932-39. The way out of the crisis cannot be shown by the Western countries and/or international organisations; they all defend the interests of multinational companies. "Apertura" will cause Turkey to fall victim into the hands of these companies. To the extent that Turkey has to export she must go into the Middle Eastern (preferably Moslem?) markets and certainly must not accept the role of the "greengrocer, butcher, milkman of the West", a role (presumably) suggested and forced upon Turkey by the international organisations.

"Export pessimism": Trade may be beneficial but only faintly so if Turkey is to export primary commodities or light consumer goods. Agricultural prices are imposed by the industrial countries; income elasticities of demand for agricultural products is low; price elasticities of agricultural suppliers is low and therefore they cannot respond to wide fluctuations; agricultural specialisation cannot provide technical progress. Light consumer goods, too, face similar conditions. Further, other export-oriented developing countries have taken the lead, so Turkey stands very little chance to enter these markets. In any case, Turkish industry is not prepared to export manufactures. On top of all these, the world has entered a new era of protectionism.

"Equity considerations": Export-oriented growth may be preferable from the point of view of the country as a whole, but this can only be done by increasing the inequalities in general and especially by reducing real industrial wages. (The assumptions here must be that Turkish industry works at full capacity and uses its intermediate inputs efficiently, and that the rates of profit are not higher than they are in competing countries.) The necessity of keeping wages low lead to political repression. Success stories of "apertura" show that rightwing dictatorships are necessary ingredients of outward-looking policies. (Here, the argument takes two different forms: one is the view that industrial capitalists have decided in favour of "apertura" so there is nothing to do under the present socio-economic framework and the other is the view that even under the present framework the contradictions between the local industrialists and foreign companies can be used to convince the former to prevent "apertura". From this point on, the second version resembles the "nationalistic" approach.

"Revolutionary strategy": "Apertura" means integration with monopoly capitalism. Inward-looking strategy is the only way to break the links with the latter. Monopoly capitalism is in direct contradiction with the interests of "national bourgeoisie" which are, in their turn, in contradiction with those of workers. Since the crisis is not of import-substitution but of capitalism, the solution to the problems of poorer classes - who suffer from the crisis more than all other groups - cannot be found by changing the external strategy. Import-substituting industries, however small additional employment they create, help to build the basis on which the "revolution" must be launched. Socialist development strategy is inward-looking; export-oriented industries create difficulties in terms of central planning. For the sake of of revolution, which will be realised by a "vanguard" consisting of revolutionary intellectual and industrial (metalworking?) proletariat [21], further import-substitution by the public sector must be promoted. If these industries will already be in the hands of government, the revolution will have an easier job to do in term of expropriation. Until that time, all that can be done is "bureaucratic guerrilla warfare" which aims at obstructing all foreign investments, even if it is coming to establish basic capital goods industries, and most private investments in order to strengthen the capitalist state (which is supposed to be taken over by the revolution).

7. "DEPENDENCE" ARGUMENT AGAINST "APERTURA"

Perhaps the most widespread argument against "apertura" in Turkey is based upon the assumption that the relationships between Turkey and the Western world create dependence. The assumption is an all-embracing statement; it covers both exports and imports, as well as financial and technological relationships. In other words, Turkey is assumed to be on the "losing" side when she increases her exports (because the prices "imposed" by the powerful buyers), when she imports more (both because of "imposed" prices and because of not being able to produce these items domestically), when she borrows (because she has to accept the conditions set by the creditors and, also, because she has to pay these loans back (!), and even when she transfers technology (because it is the technology that is surpassed already and because Turkey has to pay royalties and licence fees).

"Dependence" argument feeds on the nationalistic and traditional feelings, which are themselves partly the product of historical hostilities between European (Christian) and Ottoman (Moslem) societies, and on the modern frameworks of conflict like the dichotomies of "centre-periphery", "North-South", or Marxian theory of "imperialism". It is interesting to note that in the recent debate, attempts to appeal to a wider strata of population resulted in a mixture of nationalistic (even religious) and "North-South" type arguments (Soysal, 1979c and 1980c). [22] However, the more commonly used version of the "dependence" argument attributes "apertura" to the IMF and the World Bank coercion (Dogan, 1980; Soral, 1980; Soysal, 1978c and 1979b; Sonmez, 1980:50-51; Turel, 1979:6-7; Turkcan, 1980: 120-22).

The reasoning behind that argument must be as the following: Because Western countries see Turkey as a potential rival they do not want her to industrialise and they try to obstruct her efforts at establishing basic and "heavy" industries, using the IMF and the World Bank, which represent the "common interests of the Western world". Boratav (1979) provides the most explicit form of this "reasoning". Further, an international conspiracy - in the narrow sense "plotting behind closed doors" - is assumed (or shown). [23]

It would be naive to think that the "development conflict", rather than the idea of "partners in development", is particular to the Turkish intelligentsia. Most less developed country elites believe that the developed world, together with its multinational organisations, do not want the less developed countries to develop and actively try to slow down or to divert the development efforts of the latter. In the early post-war period, the basic issue in this "conflict" was assumed to be industrialisation; the developed world did not want the rest of the world to industrialise to keep them dependent. This assumption has had to be qualified twice since its first appearance, primarily because import-substituting-industrialisation depended, not only on the consent, but on the active cooperation of the foreign capital - in Turkey, too, foreign capital is concentrated mostly in inward-looking industries (Alpar, 1977:169ff; Uras, 1979:143ff) - it has become difficult to argue that the "centre" did not want the "periphery" to industrialise. Hence the qualification that "what is not permitted" was not "industrialisation in general" but "the development of heavy industries" or "real industrialisation". The second qualification came when, either because of a shift in protection policies or as a consequence of changing comparative advantages, heavy industry started to shift to less developed countries. It was then suggested that the "conflict" and the mechanism of dependence should be sought in "technology" (Frank, 1969:211ff) or in the exclusion of "a more restricted group of capital goods industries" (Sutcliffe, 1972:190). [24]

Note that the "development conflict" as understood by "dependencia" theorists is a "zero-sum game", i.e. the possibility that the "South" gains anything from relationship with the "North" is totally excluded, even if it gains something in economic terms it pays back in terms of political dependence (Hufbauer, 1981). Further, many authors go as far as to suggest a wider cooperation (conspiracy?) in the "North", including the Soviet Union, of course, Eastern European countries and Southern Europe, as well as the Third World proper (Galtung, 1976; Levinson, 1979). Under such circumstances, i.e. all "escape routes being blocked", "self-reliance" (if not total autarky) is promoted as the antidote, while outward-looking strategies of any kind are judged to be defeatist.

It must be noted also that the "conflict" assumed by "dependencia" is not an extension of "group interests", which are bound to exist under all circumstances. Instead, as noted earlier, the "groups", or a "group of countries", are thought to have common interests vis-a-vis another group. So they organise.

Let us now examine the validity of "dependencia" in terms of its internal consistency as well as of empirical facts relating to the Turkish experience. First of all, if the IMF and the World Bank represent the "Western interests as a whole" against less developed countries (Hayter, 1971; Payer, 1974) these interests must be uniform and harmonious, at least vis-a-vis less developed countries. Yet during the recovery from 1974-75 recession, for example, the latter were in

> "...the cross-fire of fears and hopes of bankers, industrialists and policy-makers in the industrialist countries. Those with a stake in, or responsible for the world financial system have been urging the oil importing developing countries...to tighten their belts, import less and export more. Those thinking in industrial terms have been urging the opposite." (Blackhurst, et.al., 1978, p.2)

Which are the "common Western interests" then? How are they reconciled by the IMF or the World Bank? While a fierce competition between giants of the world business is going on, (and getting fiercer under crisis conditions), it is too simplistic to conceive "Western world" and/or "its interests" as a whole. Extremely complicated trade, investment, etc. relationships of the international market between the individual firms, countries and country groups cannot be summarised in such general dichotomies in any useful manner. "North vs. South" approach may, of course, help in increasing the bargaining power of the less developed countries, but not in understanding the real issues facing the individual countries. An oil-importing less developed country, for example, must stretch her mind considerably to believe that the extreme balance-of-payments difficulties she has been facing since 1974 are the result of a "Western conspiracy" and not of the OPEC decisions. The fact that the induced price rises in industrial prices rendered the hardship greater does not save "North-South" approach unless more far-fetching conspiracies are suggested. [25]

A natural extension of the assumptions "that the Western world does not want the less developed countries to industrialise and tries to prevent it" and "that the IMF and the World Bank represent the Western interests" is a third assumption "that any proposal by these organisations is made to obstruct industrialisation efforts". Hence all such proposals must be rejected. [26]

During the recent talks with the IMF and the World Bank, the widespread attitude in Turkey was "non-cooperation" if not "hostility". This is partly explained by the unpopularity of the IMF's "austerity package", but the basic notion behind the hostile attitude was that the IMF and the World Bank were there as the representatives and negotiators "the West", i.e. "cause of all troubles", and not as multilateral organisations to help Turkey to stabilise her economy and to use her scarce resources more effectively so that she becomes financially more self-reliant and starts to grow again. [27] That conspiratorial role attributed to these organisations caused some authors to miss the point in the basic logic of stabilisation programmes. Soral, for instance, has argued that the "IMF proposals" cannot represent effective remedies for the crisis of the Turkish economy for, he maintained, they have already been tried and proved to be futile. [28] He has suggested instead "Kemalist etatism" (1980:30) - not only as a nostalgic concept, apparently, but also as a remedy to the present crisis. Unfortunately, Soral does not define what he means by "Kemalist etatism", but presumably this concept refers to the policies followed between 1932 and 1938. Yet even a brief look at the policies of that period reveals that an extreme similarity between those policies and the "IMF/World Bank proposals" (some of these proposals are attributed to these organisations).

First, Turkey maintained a positive trade balance in those years with the exception of one year (SIS, 1952:369). Secondly, Turkey did not hesitate to benefit from foreign capital, i.e. she was financially outward-looking (Kucuk, 1978:131, 141-43; Rozaliev, 1978:158-62). Thirdly, over that period, wages were kept constant (there was a 3 per cent fall in real wages) while profits increased (Bulutay, et.al., 1974:app. Tables 25, 54, 101/a; Cavdar, et.al., 1973:154-56; Hershlag, 1968:96-107). Finally, in spite of an economically active government, monetary expansion was kept within reasonable limits [29] and internal borrowing did not exceed 10-12 per cent of budget revenue (Hershlag, 1968:87). Those who are inclined to offer pre-war etatism as an alternative to economic liberalisation and "apertura" must therefore take into account not only the nostalgic and glorified elements of that period, but also the basic logic of austerity underlying the etatist trade, monetary and industrial policies, without, of course, forgetting the not-so-egalitarian distributional aspects. [30] Failing to do this, one is left in the awkward position of arguing against "apertura" and/or stabilisation measures not because they may have negative effects on distribution, on the industrial structure or on the overall rate of growth of the economy, but because they are suggested by the international financial organisations (Toplumcu Dusun, 1979: 427-28).

In fact, "dependence" version of the argument against "apertura" suffers most from the confusion of the concepts of "trade dependence" and "financial dependence". "Trade dependence" is considered to be "dangerous" for several reasons: (i) export demand is sluggish, (ii) terms of trade against the less developed countries, (iii) the raw material exporting countries do not have the flexibility to adjust themselves to changing market conditions, (iv) export-oriented economy does not contribute to the technical knowledge, or to the training of skilled administrators, managers, technicians and workers, (v) export sectors fail to provide a focus around which an integrated national economy can develop, (vi) export-oriented economy has developed patterns of distribution and economic-political power that thwart modern industrial and agricultural growth (Green and Seidman, 1968:37-51).

As is readily seen, the kind of export activity described here is colonial. Exporters have no control on the supply of commodities, they cannot use the export revenue in the other sectors of the economy, and in fact, the sector is assumed to be owned and managed by expatriates. There should be little doubt that the contribution of the export sector to the overall growth of the economy under these conditions cannot go beyond its domestic value added (net profit repatriation). Yet to base the opposition to outward-looking trade policies, as they are generally proposed and set into work today, on the characteristics of colonial trade cannot be justified. Economic dependence that may be associated with trade activity is a result of the latter's specific organisation and mechanism, and not of trade in general. At least one of the important sources of bias in trade policy today is the assumption that trade between developed countries takes place on colonial principles (Streeten, 1980:4).

One must also note that "dependence" emanating from trade is sometimes blamed on specific commodities or commodity groups. In this context, Singer (1971) is very instructive in showing the fallacy of regarding some commodities as "bad" and some "good" with the implication that exporting (or even producing) the former brings "humiliation" and the latter "dignity". Hints at a possibility that some food industries (meat and dairy products, canned foods and fresh vegetables) are likely to become competitive in the European market created an outburst of anger in Turkey, and many politicians, economists, and column-writers "declared" that Turkey "would never accept the roles of the milkman, grocer, greengrocer or the butcher of Europe". Interestingly, Turkey has been selling livestock to neighbouring Middle Eastern countries through legal and illegal channels for a very long time, but those who have opposed to the "role of butcher" had not been so sensitive to the "humiliation of being a shepherd". Irrationality in such behaviour stems from colonial-type reactions as well as from the feelings deeply embedded in the social fabric.

Trade expansion suggested in outward-looking industrialisation strategies in general and in the concrete proposals made by various authors for the development of Turkish exports has nothing in common with the colonial pattern of trade. The degree of dependence created by import-substituting-industrialisation is much greater than that created by an export-oriented strategy. One reason for this is the fact that the former leads to higher resource gap/export ratios, hence to a greater need for external finance under favourable conditions. While export revenue growth is constrained by a heavy bias against export sectors, bleak export prospects reduce the number of creditors who are willing to make commercial loans on current terms and therefore force the country to go to the "sharks" of international money market. Credit-worthiness of a country is also affected by the IMF approval of its economic programme while the stringency of measures proposed by the IMF, especially the rate of devaluation, is likely to vary directly with the size of the country's trade gap (Yotopoulos and Nugent, 1976:319).

Import-substituting-industrialisation not only deepens foreign exchange problem but also restricts policy options to cope with it. A country dependent on imports for its consumer goods may have to reduce the consumption of imported products when faced with a foreign exchange crisis, but the countries which followed an import-substitution strategy suffer, under similar shocks, capacity underutilisation and massive lay-offs in import dependent industries on top of the reduction of consumption.

Under such conditions, governments face few alternatives: first is to go to the IMF which is likely to propose politically difficult stabilisation measures without a great emphasis on long-term solutions of the foreign exchange problem. Second alternative is to seek concessional loans on a bilateral basis, but these often involve political blackmail [31] which may be difficult to meet. Governments then try the third alternative: short-term loans with very high margins from the international money market. Although bilateral loans and commercial credits are often made dependent on the approval of the IMF (the so-called "green light"), countries can "get away" with such loans if they are willing to pay extraordinarily high rates of interest or to meet the political demands made by the creditor governments. Governments trying to avoid the IMF stabilisation measures mainly for short-sighted political reasons are likely to face two serious problems: (i) higher finance cost attached to the loans obtained without the IMF approval, and (ii) the postponement of the measures that have to be taken to cure the "addiction" to external finance.

Recent developments in the financial situation of the Turkish economy illustrate the dangers of a short-sighted policy to cope with long-term problems. Failing to identify the direct and indirect costs of import-substituting industries on the Turkish economy, successive governments tried to keep the "traditional" strategy intact even under the new circumstances brought about by the oil crisis. Thus during the period 1975-78 Turkish short-term borrowing reached $3.6 billion worth of very expensive loans under "convertible lira accounts" or similar schemes. [32] The IMF has already warned that the policy of short-term borrowing at very high margins was very dangerous, and the prompt repayment of these loans was very unlikely. When "extravaganza" had to come to an end and Turkey attempted to reschedule her debt (following the stand-by agreement with the IMF in March 1978) banks responded positively in order to avoid massive write-offs. But in spite of the IMF "green light" they were not willing to provide fresh loans. "Green light" was only a necessary condition and was not sufficient to mobilise the financial institutions when they remained suspicious of the credit-worthiness of Turkey. [33] In spite of the agreement with the IMF on the stabilisation programme, the Turkish government was very reluctant to effect the measures stipulated in its letter of intent (IMF, 1979; Olgun, 1979). "Aid-addiction" and laxity had reached such great proportions in the Turkish economy that the success or failure of a government was being assessed in its ability to find further loans and the "measures" were being taken (or written in the documents) to obtain the latter. The whole economy waited (or wasted) two years for these loans that were supposed to "save the economy from its troubles", while no real effort was spent to reorganise for an export drive or for a more efficient allocation and use of resources. The strategy of import-substitution has almost paralysed economic decision-making and led the country into the most humiliating form of dependence.

All these do not mean that outward-looking framework excludes the use of foreign financial resources. The critical difference between inward and outward-looking strategies in this respect lies in the employment of these extra resources. While inward-looking strategy attempts to use these resources for the development of so-called "foreign exchange saving" industries, external finance is used to promote "foreign exchange earning" industries in outward-looking strategy. At the first glance, they seem to have the same effect on the balance of payments, but the development of these two groups of industries follow very different paths in terms of employment and income distribution, of industrial structure and X-efficiency, of the level and form of government intervention, of their impact upon the general outlook of the society (Keesing, 1967). There are two points related to "dependence" argument which need to be stressed.

First, outward-looking strategy unlike import-substitution, forces the economic decision-making mechanism to be very flexible and responding to the external opportunities and shocks. Secondly, the attitude of the world financial circles towards outward-looking economies is much more favourable. The latter point is often attributed to the authoritarian regimes in some outward-looking countries. While this interpretation may be plausible in some cases, it hardly offers an explanation for the general pattern. There exist many inward-looking dictatorships which do not get a favourable treatment in the world of finance, and many outward-looking democratic regimes which are able to use very large sums of foreign credit. Confidence is restored more by a growing export revenue (hence a lower risk) than by dictators (who, in some countries, change even more frequently than democratic governments).

To sum up, outward-looking strategy depends on building a mechanism which will effectively make use of international capital, especially when the latter is relatively cheaper. An economy that borrows, pays back promptly and invites equity capital is financially independent so long as the terms of such transactions are not dictated by the lenders. On the other hand, this country may choose not to produce every single item it consumes, perhaps not even the goods she may need for her growth or for her defence, but so long as she is able to participate fully in the world trade according to her comparative advantages, "trade dependence" will not be a liability but an asset, increasing the overall efficiency and flexibility of the economy.

NOTES

1. It must be pointed out that even when a country is outward-looking in technological terms, an indigenous research and development capacity to adapt to foreign technology is needed, given the problems of soil,

materials, environment, skill and climate. See Freeman, 1974, p.280.

2. For various interpretations of Soviet foreign trade behaviour, see Ebiri, 1975, pp.216-22; Herman, 1698; Wilber, 1969, pp.103-8, and for the transfer of technology, see, Berliner, 1966; Campbell, 1960, pp.164-68, and Carr-Davies, 1969/I, pp.441ff.

3. For Japanese policy on technology transfer, see Blumenthal, 1976; Macrae, 1967.

4. For alternative forms of measurement, see Chenery, 1960; Maizels, 1963 and 1970, pp.149-51; Morley-Smith, 1970. An application of import/supply ratio measurements to the Turkish economy is provided by Korum, 1977. Chenery, 1980, uses the "import ratio" approach to identify the role of trade strategies on industrialisation. De Pablo, 1977, too, attempts to use the change in the "import ratio" to identify import-substitution. He observes that the term "import-substitution" is used in the literature both to indicate a change in the composition of imports (which has always existed and will exist in the future) and to express a reduction in the ratio of imports to gross domestic product (which may not be a continuous trend).

5. One must take into account, however, the possibility of having to import auxiliary inputs and packing materials.

6. See, for instance, Baer and Samuelson, 1977; Little and Scitovsky, 1970; Song, 1973. Also see, Bhagwati, 1978, p.207, and Desai 1969, for similar objections.

7. See, for instance, Chenery, 1975, p.314; Mitra, 1974; Pomfret, 1975, and Robock, 1970.

8. It is interesting to note that one of the early proponents of this view is Turgut Ozal. See, his contribution to a panel met to discuss long-term prospects of the Turkish economy in Ustunel, et.al., 1975.

9. However, Birand (1979, pp.513/529/534) shows that to interpret the "uncooperative" attitudes of the multinational banks and of Western governments as an extension of their efforts to force Turkey into an agreement with Greece over Cyprus is not justified.

10. As is well known, there is more than one version of Marxian theory of imperialism, and some of these reach diametrically opposite conclusions on the effects of industrial growth in the "centre" on the "periphery", or of the crisis of the world economy. See Barratt-Brown, 1974; Frank, 1966; Magdoff, 1969; McMichael, et.al., 19741; O'Connor, 1970; Warren, 1973 for recent examples.

11. Keyder (1979) does not accept the traditional "dependencia" argument but argues that the crisis of the Turkish economy is part of the world economic crisis.

12. It is never made explicit just what was expected of the IMF in 1977 and later. Presumably, the IMF in order not to be a part of international conspiracy, should have poured financial support with "no questions asked" so that the over-valued exchange rate, subsidised

oil, huge public sector losses, uncontrolled money expansion together with a rapidly rising debt/service ratio, could be maintained. In other words, Turkey should continue with "business as usual" as if the economy was smoothly growing.

13. Lausanne Treaty, signed in 1923 between the Turkish government and the Allied Powers, required the Turkish government not to impose higher tariffs (than those effective in 1916) and other trade restrictions. However, "escape clauses" were so flexible (Hershlag, 1968, p.22) that the Treaty could have hardly restricted government action towards the protection of local industry.

14. Post-1965 performance is used as "evidence" also for "the relative effectiveness of planning over an unplanned economy" or for "the relative superiority of Justice Party governments over others in promoting economic growth".

15. A similar comparison, though with somewhat different data conclusions, can be found in Hatiboglu, 1978, pp.209-11.

16. The planners had still not realised the shortcomings of that "method" to distribute import-substitution "tasks" between industries without regard to allocative efficiency in the preparation of the Fourth Five-Year Plan, i.e. in the midst of the heaviest foreign exchange crisis of the country. See SPO/DEP/LTPD, 1977, pp.11-17.

17. See Ebiri (1975, pp.140ff) and SPO/DSP (1977, pp.52-65) for the methodology of calculation. The share of capital goods investment in the total planned investment is 16.1 per cent in the Second Plan and 17.6 per cent in the Third Plan. These are not the shares of the "engineering industries", which are inappropriately christened by the Third and Fourth Plans as "capital goods industries", but are calculated by weighting investments with their direct and indirect contributions to gross fixed capital formation, excluding residential construction.

18. Fourth Plan allocates only 15.5 per cent of nonresidential investment to capital goods sectors. This is a share lower than Second Plan's planned and actual figures, too. (State Planning Organisation has not published the sectoral breakdown of manufacturing investments since 1976. It is not possible, therefore, to compare the Fourth Plan targets with the actual allocation of investments in the Third Plan period.)

19. See Baer and Samuelson (1977, pp.1-2) for the similar problems created by import-substitution in Latin American countries. Hatipoglu (1978, p.142ff) and Akalin (1980, pp.184-85) point to the problems of the traditional development strategy in Turkey without making the links explicit.

20. A very interesting example of the import-substituting industrialist attitude is provided by Vehbi Koc, one of the few most influential businessmen in Turkey, in an interview (Ipekci-Koc, 1976). Put under pressure by the interviewer on the question of what his family of corporations, Koc Holding, does to earn the foreign exchange that is required for intermediate and capital inputs, he argues that these

corporations cannot do anything "because exporting is a completely different activity". He tries to justify this instead by arguing that Koc Holding "saves a lot of foreign exchange by producing Fiat cars in the country". Warned about the possibility of foreign exchange crisis becoming more severe in the coming period and asked what will happen to their factories when they cannot find the inputs that have to be imported, Koc brings forward "the danger of massive lay-offs", presumably to raise public attention to get further privileges in the allocation of scarce foreign exchange. At the same time, however, Koc seems to have been convinced that "the Common Market entry will be beneficial" and export processing zones and more foreign capital in general must be encouraged.

21. The "revolutionaries" do not seem to be very interested in the size and growth of industrial labour force. When confronted with the superior performance of export-promoting strategies in terms of industrial job creation through relatively more labour-intensive techniques, their response is usually to point to the possibility of declining income difficulties between the present workers in large industrial establishments and the poorer groups (the unemployed, agricultural workers, younger workers in small industry, those working in marginal jobs, etc). The latter group's chance of being employed by export-oriented industries cannot, in my view,be neglected by the working class parties in the developing countries unless, of course, they adopt narrow "unionist" attitudes.

22. Kucuk (1980, pp.546-49) however argued that "imperialism was hatching an Islamic religious regime in Turkey", probably "through a military coup d'etat". (The military intervened on 12th September, 1980 - almost simultaneously with the publication of Kucuk's book - and among the few immediate objectives of the generals was "to eradicate the teocratic movements"!)

23. Like the IMF openly approving Turkish government's "letter of intent" in April 1978 but "leaking" to the commercial banks of its disapproval so as to hold them back from rescheduling Turkey's debt, (Dogan, 1980, p.150).

24. The majority of Turkish authors who employ the "dependence" framework in their analysis do not seem to have reached the last stage yet. See, for instance, Boratav, 1979; Deniz and Emre, 1980; Ozgur, 1976; Somer, 1979; Sonmez, 1980, pp.11-13; TIP, 1978, pp.15-20, 55. However, see also Gulalp, 1979; Keyder, 1979 and Seni, 1978, as rare examples of minority, but by no means uniform views.

25. Rothstein (1977, p.14ff) points out the effects of the OPEC movement on the other less developed countries and seeks answers to the question why OPEC victims have not protested.

26. Even when these proposals represent the only way out of crisis and have already been made independently by the government officials or by academic circles. Most prominent examples of this obsessive opposition are seen in the discussions relating to import-substituting consumer goods industries, which were given the derogatory name of "assembly industries" since mid-sixties, deficit financing, low interest rates, convertible lira accounts, agricultural support

policies. These issues were traditionally used to criticise governments, but when they take place in a "letter of intent" or in an IMF/World Bank document, they are considered "points of coercion". (Soysal, 1978a and 1980b) provide a very illustrative comparison; in the former, the author points out to the futility of price controls, but in the latter, he suggests that since the country is under military discipline, the economy should not be left to the "chaotic" price mechanism, especially when the latter is part of and IMF/World Bank inspired liberalisation programme.

27. The role of "negotiator" attributed to these organisations sometimes included political issues, too. See Dogan, 1980, p.140ff; Soysal, 1978b, 1979a and 1980a, as examples of interpreting delayed loans in terms of Cyprus problem.

28. It is very common to think only of the devaluation of currency as the "IMF prescription". Soral must have in mind devaluation only when he writes about "usual" IMF proposals. In this context it may be useful to remind the reader that the IMF does not always suggest devaluation. In many cases it opposed devaluation. See Higgins, 1968, p.290.

29. "The government made efforts to allow an increase in currency in circulation only to the extent that the reserves of the Central Bank in gold and foreign currency expanded, as a measure against the temptation to indulge in excessive money printing". Hershlag, 1968, p.85.

30. For a summary statement of the effect on income distribution of Turkish etatism, see Boratav, 1963, pp.75-76.

31. See Olson, 1979, on the politics of "import-substitution" path that led Brazil to aid-addiction and on the manipulation of foreign aid by the donors to affect internal politics of Brazil.

32. See World Bank, 1979, p.31. Most of these loans were for very short-term (around one year). Although a limit of 1.75 percent on the margin over LIBOR was imposed by the Turkish government, most banks were able to obtain 5 to 10 per cent commissions from the local businessmen. Economist, 1978b, pp.120-21.

33. Spreads were very low in the European market and the borrowers in general faced very favourable conditions in the first half of 1978, when Turkey failed to obtain fresh loans. Economist, 1978a, pp.6-15.

REFERENCES

Akalin, G. (1980); "Ekonomik Buhran ve Partiler" (Economic Crisis and the Political Parties), Toplumcu Dusun, No.16, pp.182-87

Akgul, N. (1976); "Turkiye'de Montaj Sanayii" (Assembly Industries in Turkey), 1976 Conference on Industry, Chamber of Commerce, Ankara

Alpar, C. (1977); Cok Uluslu Sirketler ve Ekonomik Kalkinma (Multinational Corporations and Economic Development), Academy of Economics and Commerce, Ankara

Altintas, M. (1978); Ice Yonelik Sanayilesme Politikasi (Inward-looking Industrialisation Policy), Mugla Academy of Economics and Commerce, Ankara

Baer, W. and Samuelson, L. (1977); "Editors' Introduction", World Development, No.5, pp.1-6

Balassa, B. (1977); "Export Incentives and Export Performance in Developing Countries: A Comparative Analysis", World Bank Staff Working Paper, No.248, World Bank, Washington D.C.

Barratt-Brown, M. (1974); The Economics of Imperialism, Penguin, Harmandsworth

Berliner, J. (1966); "The Economics of Overtaking and Surpassing" in H. Rosovsky (ed.), Industrialisation in Two Systems: Essays in Honour of Alexander Gerschenkron, Wiley, New York

Berry, A. and Thoumi, F. (1977); "Import-Substitution and Beyond: Colombia", World Development, No.5, pp.89-109

Bhagwati, J. (1978); Anatomy and Consequences of Exchange Control Regimes, National Bureau of Economic Research, Balinger, Cambridge, Mass.

Birand, M. A. (1979); Diyet (Ransom), Milliyet, Istanbul

Blackhurst, R., et.al. (1978); Adjustment, Trade and Growth in Developed and Developing Countries, GATT Studies in International Trade, GATT, Geneva

Blumenthal, T. (1976); "Japan's Technological Strategy", Journal of Development Economics, No.3, pp.244-55

Boratav, K. (1963); Turkiye'de Devletcilik (Etatism in Turkey) Istanbul

Boratav, K. (1979); "IMF'nin Niyetleri ve Ithal Ikamesi" (Import-Substitution and the IMF's Intentions), Milliyet, August 27

Bortucene, I. (1979); "Ekonomik Onlemler Uzerine Bazi Dusunceler", (Some Thoughts on the Stabilisation Programme), Yeni Ulke, No.7, pp.11-40

Bulutay, T. et.al., (1974); Turkiye Milli Geliri (National Income of Turkey), Faculty of Political Science, Ankara

Campbell, R.I. (1960); Soviet Economic Power: Its Organisation, Growth and Challenge, Houghton-Mifflin, Cambridge, Mass.

Carr, E.H. and Davies, R.W. (1969); Foundations of a Planned Economy I: 1926-1929, Penguin, Harmandsworth

Central Bank of Turkey (1980); 1979 Yillik Rapor (Annual Report: 1979), Central Bank, Ankara

Chenery, H.B. (1960); "Patterns of Industrial Growth", American Economic Review, Vol.50, 624-53

Chenery, H.B. (1975); "Structuralist Approach to Development Policy", American Economic Review, Vol.65, pp.310-16

Chenery, H.B. (1980); "Interactions Between Industrialisation and Exports", American Economic Review, Vol.70, pp.281-87

Cavdar, T., et.al. (1973); Turkiye'de Toplumsal ve Ekonomik Gelismenin 50 Yili (Fifty Years of Social and Economic Development in Turkey), State Institute of Statistics, Ankara

Cetin, H. (1972); "Perspektif Plan Uzerine Konusma" (Speech on Perspective Plan) in Ucuncu Plan Modeli Uluslararasi Kollokyomu (International Conference on the Third Plan Model), February 28-29, State Planning Organisation, Ankara

Demirer, M.A. (1979); "Seksenli Yillarda Turkiye" (Turkey in the Eighties), Forum, No.3, October 15, pp.15-16

Deniz, M. and Emre, G. (1980); "Gunumuz Turkiye'sinde Demir-Celik Sorunu Uzerine" (On Iron and Steel Problem in Today's Turkey), Yeni Ulke, No.10, pp.191-221

De Pablo, J.C. (1977); "Beyond Import-Substitution: The Case of Argentina", World Development, No.5, pp.7-17

Dervis, K. (1974); "Doviz Birikimi, Kambiyo ve Dis Ticaret Politikalari" (Trade and Foreign Exchange Policies and Reserve Accumulation), Banka ve Ekonomik Yorumlar, Vol.11, No.1, January, pp. 26-35

Dervis, K. and Robertson, S. (1978); "The Foreign Exchange Gap, Growth and Industrial Strategy in Turkey", World Bank Staff Working Papers, No.306, World Bank, Washington D.C.

Desai, P. (1969); "Alternative Measures of Import-Substitution", Oxford Economic Papers, No.21, pp.312-24

Divitcioglu, S. (1980); "M. Barlas ile Gorusme" (Interview with M. Barlas), Milliyet, February 18

Dogan, Y. (1980); IMF Kiskacinda Turkiye, 1946-1980 (Turkiye in the Stranglehold of the IMF, 1946-1980), Toplum, Ankara

Ebiri, K. (1975): An Outward-Looking Socialist Development Strategy with Specific Reference to the Turkish Economy, unpublished D.Phil. Thesis University of York (G.B.)

Ebiri, K. (1979a); "Introduction and a Summary of Main Findings" in K. Ebiri, G. Tuzun and Y. Kepenek (eds.), Growth, and Development of the Turkish Manufacturing Industry, Economic and Social Research Institute, Middle East Technical University, Ankara

Ebiri, K. (1979b); "Niyet Mektuplari ve Turkiye'nin Niyetleri" (Letters of Intent and Turkey's Intentions) Milliyet, August 18

Economist, The (1978a); "Must Lend, Will Travel - International Banking: A Survey", March 4

Economist, The (1978b); "Hobson's Choice: Turkey's Debt", October 28

Erdem, T. (1980); 80'leri Karsilarken (Meeting the Eighties), Ajans Turk, Ankara

Erdogdu, V. (1978); "Emperyalizmin Guncel Celiskileri ve Turkiye Kapitalizminin Bunalimi" (Actual Contradictions of Imperialism and the Crisis of Turkish Capitalism), Yeni Ulke, No.2, pp.5-72

Erdogdu, V. (1979); "Turkiye Kapitalizminin Agirlasan Toplumsal ve Ekonomik Sorunlari" (Aggravating Social and Economic Problems of Turkish Capitalism), Yeni Ulke, No.8, pp.17-40

Frank, A.G. (1966); "Underdevelopment of Underdevelopment", Monthly Review, Vol.8, No.4

Frank, A.G. (1969); Capitalism and Underdevelopment in Latin America, Monthly Review, New York

Frankena, M (1974); "The Industrial and Trade Control Regime and Product Designs in India", Economic Development and Social Change, Vol.22, pp.249-64

Freeman, C. (1974); The Economics of Industrial Innovation, Penguin, Harmandsworth

Galip, A. (1979); "Turkiye Kapitalizminin Krizi Uzerine" (On the Crisis of Turkish Capitalism), Birikim, No.54-55, pp.78-85

Galtung, J. (1976); "Conflict on a Global Scale: Social Imperialism and Sub-Imperialism - Continuities in the Structural Theory of Imperialism", World Development, No.6, pp.153-65

Green, R. and Seidman, A. (1968); Unity or Poverty? The Economics of Pan-Africanism, Penguin, Harmandsworth

Gulalp, H. (1979); Yeni Emperyalizm Teorilerinin Elestirisi (Critique of the New Theories of Imperialism), Birikim, Istanbul

Gulalp, H. (1980); "Turkiye'de Ithal Ikamesi Bunalimi ve Disa Acilma" (The Crisis of Import-Substitution in Turkey and Outward Orientation), METU Studies in Development, Vol.7, No.1-2, pp.37-66

Hatipoglu, Z. (1973); "Ekonomimiz En Kuvvetli Asamasindadir" (Our Economy is at its Strongest Juncture), Banka ve Ekonomik Yorumlar, Vol.10, No.12, December, pp.9-10

Hatipoglu, Z. (1978); An Unconventional Analysis of Turkish Economy: An Essay on Economic Development, Aktif Buro, Istanbul

Hayter, T. (1971); Aid as Imperialism, Penguin, Harmandsworth

Herman, I. (1968); "The Promise of Economic Self-Sufficiency Under Soviet Socialism" in V.G. Tremi and R. Farrell (eds.), The Development of the Soviet Economy: Plan and Performance, Praeger, New York

Hershlag, Z.Y. (1968); Turkey: The Challenge of Growth, Brill, Leiden

Higgins, B. (1968); Economic Development, Second Edition, W.W. Norton, New York

Hufbauer, G.C. (1970); "The Impact of National Characteristics and Technology on the Commodity Composition of Trade in Manufactured Goods" in R. Vernon (ed.), The Technology Factor in International Trade, National Bureau of Economic Research, Columbia University Press, New York

Hufbauer, G.C. (1981); "Corporate Investments", interview given to J.J. Harter in V.S. Kreuetzer (ed.), Views on Global Economic Development, U.S.I.C.A, Washington D.C.

Ilkin, S. (1978); "Devletcilik Doneminin Ilk Yillarinda Isci Sorununa Yaklasim ve 1932 Is Kanunu Tasarisi" (Attitudes Towards Labour Problem in the Early Etatist Period and the Labour Legislation Proposed In 1932), METU Studies in Development, Special Edition, 251-348

IMF (1979); "Turkey: Staff Report for the 1978 Article IV Consultation", February 8

Ipekci, A. and Koc, V. (1976); "Vehbi Koc" Milliyet, May 31 and June 1

Ipekci, A. and Ustunel, B. (1973); "1971'den 1973'e" (From 1971 to 1973), Milliyet, December 3

Kafaoglu, Arslan B. (1979); Enflasyon: Gelismis ve Azgelismis Ulkelerde (Inflation in Developed and Underdeveloped Countries), Tekin, Istanbul

Kandiller, R. (1976); "Dis Ticaret Acigi Sorunu ve Gorunmeyen Kalemlerdeki Gelismeler" (Trade Gap and the Changes in Invisibles), paper presented at the Conference on the Structural Problems of the Balance of Payments, February 4-5, Union of Chambers of Trade and Industry, Ankara

Kayra, C. (1972); Turkiye'nin Dis Odemeler Dengesi Tahminleri Uzerine Dusunceler (Some Thoughts on the Estimation of Balance of Payments in Turkey), Institute of Public Administration for Turkey and the Middle East, Ankara

Keesing, D.B. (1967); "Outward-Looking Policies and Economic Development", Economic Journal, Vol.77, pp.303-320

Keesing, D.B. (1979); "Trade Policy for Developing Countries", World Bank Staff Working Paper, No.353, World Bank, Washington D.C.

Keyder, C. (1979); Emperyalizm, Azgelismislik ve Turkiye (Imperialism, Underdevelopment and Turkey), Second Edition, Birikim, Istanbul

Kilickaya, E. and Ibrahimhakkioglu, N. (1978); Turkiye'nin Ihracat Yapisi (Turkish Export Structure), Centre for Export Promotion, Ankara

Korum, U. (1976); "Import-Substitution in Turkey in Relation to Integration with the EEC" in O. Okyar and O. Aktan (eds.), Economic Relations Between Turkey and the EEC, Hacettepe Institute of Economic and Social Research on Turkey and the Middle East, Ankara

Korum, U. (1977); Turkiye Imalat Sanayii ve Ithal Ikamesi: Bir Degerlendirme (Turkish Manufacturing Industry and Import-Substitution: An Assessment), Faculty of Political Science, Ankara

Krueger, A. (1980); "Trade Policy as an Input to Development", American Economic Review, Vol.70, pp.288-92

Kurdas, K. (1979); Ekonomik Politikada Bilim ve Sagduyu (Science and Sensibility in Economic Policy), Ekonomik ve Sosyal Yayinlar

Kurmus, O. (1978); "1916 ve 1929 Gumruk Tarifeleri Uzerine Bazi Gozlemler" (Some Observations on 1916 and 1929 Customs Tariffs), METU Studies in Development, Special Edition, pp.182-209

Kucuk, Y. (1978); Turkiye Uzerine Tezler (Theses on Turkey), Vol.I, Tekin, Istanbul

Kucuk, Y. (1980); Bir Yeni Cumhuriyet Icin (For a New Republic), Tekin, Istanbul

Leibenstein, H. (1957); Economic Backwardness and Economic Growth, Wiley, New York

Levinson, C. (1979); Votka-Cola (Vodka-Cola), translated by Y. Zeytinoglu, Hurriyet, Istanbul

Little, I. Scitovsky, T. and Scott, M. (1970); Industry and Trade in Some Developing Countries, OECD, Oxford University Press, London

Macrae, N. (1967); "The Risen Sun", The Economist, May 27 and June 3

Maddison, A. (1969); Economic Growth in Japan and the USSR, Allen and Unwin, London

Magdoff, H. (1969); The Age of Imperialism, Monthly Review, New York

Maizels, A. (1963/1970); Growth and Trade, (Abridged) N.I.E.S.R. University Press, Cambridge

Malkoc, S. et al. (1973); 50 Yilda Turk Sanayi (Fifty Years of Turkish Industry), Ministry of Industry and Technology, Ankara

McMichael, P. et al. (1974); "Imperialism and the Contradictions of Development", New Left Review, No.85, pp.83-104

Mitra, P.K. (1974); "Import Substitution and Export Promotion as a Means to Industrialization", Economia Internazionale, Vol.27, pp.524-37

Morley, G.A. and Smith, G.W. (1970); "On the Measurement of Import-Substitution", American Economic Review, Vol.60, pp.728-35

O'Connor, J. (1970); "The Meaning of Economic Imperialism" in R. Rhodes (ed.), Imperialism and Underdevelopment, Monthly Review, New York

Okyar, O. (1976); "Turkish Industrialization Strategies, the Plan Model and the E.E.C." in the same book as Korum (1976), pp.14-53

Olgun, H. (1979); "1979 Yilinda Turk Ekonomisinde Ana Gelismeler" (Main Trends in the Turkish Economy 1979), METU Studies in Development, Vol.22-23, pp.165-99

Olson, R.S. (1979); "Expropriation and Economic Coercion in World Politics: A Retrospective Look at Brazil in the 1960's", Journal of Developing Areas, Vol.13, pp.247-62

Oshima, H.T. (1973); "Research and Development and Economic Growth in Japan" in B.R. Williams (ed.), Science and Technology in Economic Growth, I.E.A. Macmillan, London

Ozawa, T. (1968); "Imitation, Innovation and Japanese Exports" in P.B. Kenen and R. Lawrence (eds.), Open Economy, Columbia University Press, New York

Ozgur, O. (1976); Sanayilesme ve Turkiye (Industrialisation and Turkey), Gercek, Istanbul

Ozkol, S. (1969); Geri Biraktirilmis Turkiye (Turkey Underdeveloped [by Imperialism]), Ant, Istanbul

Ozkol, S. (1970); Emperyalizm, Tekelci Kapital ve Turkiye (Imperialism, Monopoly Capitalism and Turkey), Ant, Istanbul

Payer, C. (1974); The Debt Trap: The IMF and the Third World, Monthly Review, New York

Pomfret, R. (1975); "Some Interrelationships between Import-Substitution and Export-Promotion in a Small Economy", Weltwirtschaftliches Archiv Vol.111, pp.714-27

Robock, S.H. (1970); "Industrialisation Through Import-Substitution or Export industries: A False Dichotomy" in J.W. Markham and G.F. Papanek (eds.), Industrial Organisation and Economic Development, in honour of E.S. Mason, Houghton-Mifflin, Boston

Rothstein, R. (1977); The Weak in the World of Strong, Columbia University Press, New York

Rozaliev, Y. (1978); Turkiye'de Kapitalizmin Gelisme Ozellikleri (Developmental Characteristics of Capitalism in Turkey), translated by A. Yaran, (Russian Edition published 1962) Onur, Ankara

S.I.S. (1952) Istatistik Yilligi (Statistical Yearbook), State Institute of Statistics, Ankara

S.I.S. (1978); Turkiye Milli Geliri: 1962-1977 (National Income of Turkey, 1962-1977), State Institute of Statistics, Ankara

Singer, H.W. (1971); "The Distribution of Gains from Trade and Investment Revisited", Journal of Development Studies, Vol.11, pp.376-382

Somer, K. (1979); DISK'in Ekonomik Raporu (Economic Report by the Confederation of Revolutionary Trade Unions), Yeni Ulke, Vol.9, pp.228-249

Song, In Sang (1973); "Pros and Cons of Protection and Import-Substitution", paper presented at the Rehovot Conference on Economic Growth in Developing Countries, Israel, Sept. 5-11

Soral, E. (1980); "Yol Ayrimindaki Turk Ekonomisi" (The Turkish Economy at the Crossroads), Ekonomik Yaklasim, Vol.1, No.2, pp.1-30

Soysal, M. (1978a); "Zehir Hafiye Korsan Ekonomiye Karsi" (Master Detective vs. Pirate Economy), Milliyet, March 18

Soysal, M. (1978b); "Bastirilan Turkiye" (Turkey Suppressed), Milliyet, November 21

Soysal, M. (1978c); "Hepsi Pusuya Yatmis Bekliyorlar" (They are All Waiting in Ambush), Milliyet, December 13

Soysal, M. (1979a); "Bocalama" (Confusion), Milliyet, May 5

Soysal, M. (1979b); "Bunalimdan Buyumeye" (From Crisis to Growth), Milliyet, October 17

Soysal, M. (1979c); "Cagdas Hacilar ve Papa" (The Pope and the Modern Crusaders), Milliyet, November 28

Sonmez, M. (1980); Turkiye Ekonomisinde Bunalim: 24 Ocak Kararlari ve Sonrasi (Crisis in the Turkish Economy: 24 January Decrees and their Aftermath), Belge, Istanbul

SPO (1977); Dorduncu Plan Modeli ve Cozumlerine Yoneltilen Elestri ve Onerilerin Degerlendirilmesi (Reply to the Critics of the Fourth Five-Year Plan and Its Solutions), State Planning Organisation, Ankara

SPO (1977); Dorduncu Bes Yillik Kalkinma Plani Cozumleri Uzerine Elestri ve Oneriler (A Critique of Fourth Five-Year Plan Solutions and Alternatives), State Planning Organisation, Ankara

Streeten, P. (1980); "The Choice Before Us", International Development Review, Nos.2-3, pp.3-10

Sutcliffe, R. (1972); "Imperialism and Industrialisation in the Third World", in R. Owen and R.B. Sutcliffe (eds.), Studies in the Theory of Imperialism, Longman, London

Seni, N. (1978); Emperyalist Sistemde "Kontrol Sanayii" ve Eregli Demir-Celik ("Control Industries" in the Imperialist System and Eregli Iron and Steel [Company]), Birikim, Istanbul

Tekeli, I and Ilkin, S. (1977); 1929 Dunya Buhraninda Turkiye'nin Iktisadi Politika Arayislari (Turkey's Quest for Economic Policy During the 1929 World Crisis), Middle East Technical University, Ankara

TIP (Turkiye Isci Partisi - Turkish Workers' Party) (1978); Demokratiklesme Icin Plan 1978-1982 (Plan for Democracy 1978-1982), Turkish Workers' Party, Istanbul

Togay, D. (1974); "Ihracatin Yapisi ve Ozellikleri" (The Structure and Characteristics of Exports), paper presented to the symposium on the Development Prospects of Industry and Exports, May 20-25, Centre for Export Promotion, Ankara

Toplumcu Dusun (1979); "Turkiye ve IMF" (Turkey and the IMF), Toplumcu Dusun, No.11, pp.420-30

Turel, O. (1972); "Modelin Cogaltan Katsayilari" (Multipliers of the [Third Five-Year Plan] Model), in the same book as Cetin (1972), pp.85-92

Turkcan, E. (1978); "Yeni Plan ve Turkiye" (New Plan and Turkey), Dunya, October 23

Turkcan, E. (1980); "Yeni Uluslararasi Para Sistemi Karsisinda Gelisen Ulkeler" (Developing Countries in the Face of New International Monetary System), Ekonomik Yaklasim, No.1, pp.91-124

Tuzun, G. (1979); "Burjuvazinin Niyetleri Uzerine Dusunceler" (Thoughts on the Intentions of the Bourgeoisie), Yurt ve Dunya, No.18, pp.344-58

United Nations (1978); Statistical Yearbook, United Nations, New York

Uras, G. (1979); Turkiye'de Yabanci Sermaye Yatirimlari (Foreign Investments in Turkey), Iktisadi Yayinlar, Istanbul

Ustunel, B., et.al. (1975); "Turkiye Gelecek 10 Yil Icin Ne Yapmali?" (What must Turkey do for the Next Ten Years?), a panel discussion with B. Ustunel, T. Ozal, I. Ongut, chaired by A. Gevgili, Milliyet, May 25

Varlier, O. (1972); "Ihracat Projeksiyonlari Ithal Ikamesi" (Export Projections and Import-Substitution), in the same book as Cetin (1972), pp.65-68

Warren, B. (1973); "Imperialism and Capitalist Industrialisation", New Left Review, No.81, pp.3-44

Westphal, I. and Kim, K.S. (1977); "Industrial Policy and Development", World Bank Staff Working Paper, No.263, World Bank, Washington D.C.

Wilber, C.K. (1969); The Soviet Model and the Underdeveloped Countries, University of North Carolina Press, Chapel Hill

Wolf, M. et.al. (1977); India: Export Performance, Problems Policies and Prospects, World Bank, Washington D.C.

World Bank (1979); Turkey: Policies and Prospects for Growth, World Bank, Washington D.C.

Yalin, N. (1980); "Onlem Plani ve Bazi Yeni Gorusler Uzerine" (On [Stabilisation] Measures and Some New Ideas), Yurt ve Dunya, No.19, pp.22-32

Yotopoulos, P. and Nugent, J. (1976); Economics of Development: Empirical Investigations, Harper & Row, New York

5 Short-term stabilisation policies in a developing economy: the Turkish experience in 1980 in long-term perspective*

FIKRET SENSES

1. INTRODUCTION

The pattern of trade and development in Turkey has borne a close resemblance to that of other developing countries which have relied on import-substituting-industrialisation through protectionist policies. As elsewhere, this strategy, while largely successful in the relatively easy phase of creating an industrial base in consumption goods, has, after the mid-sixties, found it increasingly difficult to extend import-substitution into intermediate and capital goods as much of domestic production and investment became crucially dependent on the availability of imports. Successive increases in oil prices in the 1970s were instrumental in aggravating the difficulties in the balance of payments which were accompanied by an acceleration in the rate of inflation, particularly towards the end of the decade. After two half-hearted attempts at stabilisation in 1978 and 1979, a major stabilisation programme, hailed as a major break away from previous policies, was announced in January 1980 amidst growing social and political turmoil. As the controversy surrounding it was in its brunt, this programme received a major boost following its adoption by the new regime, established after the military intervention in September of that year.

The main objective of this paper is to give an overview of the economic policies making up the 1980 stabilisation package and assess their initial impact on the economy. In an effort to evaluate the package within the context of long-term development goals and experience, however, we shall look beyond the short-term implications of the package and provide some insights for the medium and long-term prospects of the economy. In this attempt, we shall in Section 2 give a brief account of the overall impact of economic policies implemented as part of the import-substituting strategy in long-term perspective. This will be

* This is an edited version of the article first published in METU Studies in Development, Vol.8, Nos.1-2, 1981.

followed in Section 3 by a brief assessment of more recent developments with particular emphasis on the economy's response to successive oil shocks and attempts at stabilisation prior to the 1980 stabilisation programme. In Section 4, we shall outline the objectives and nature of main policies comprising the 1980 programme and discuss the extent to which they represent a break away from the long-term strategy. Section 5 will be devoted to the assessment of the main effects of the programme on selected indicators such as national income, investment, balance of payments, prices as well as on employment and the distribution of income. In Section 6, finally, we shall summarise our results and express our views on the medium and long-term development prospects of the economy.

2. AN OVERVIEW OF ECONOMIC POLICIES BEFORE THE OIL CRISIS [1]

The emergence of Turkey in the early 1920s as a modern nation-state corresponded with the adoption of industrialisation as the overriding national objective aimed at extending political independence to the economic sphere and catching up with the industrialised states of the West. After a brief period during which the government gave active encouragement to the development of private enterprise, the State, with the onset of the Great Depression, took a firmer grip of economic life and emerged as the chief entrepreneur in the industrialisation effort. Import-substitution in consumer goods under heavy protection induced by commercial and exchange policies and envisaging in the first place the domestic utilisation of agricultural raw materials lay at the heart of the industrialisation strategy. The world shortage of capital goods and raw materials which followed the outbreak of the Second World War was further instrumental in reinforcing the protectionist import-substitution strategy which was conducted under a strong sense of economic nationalism, resulting in the rapid expansion of manufacturing production in a predominantly agricultural economy.

The emergence in 1950 of a new political leadership in response to the grievances of private sector in general and agriculture in particular was associated with several important changes in development policy. The active encouragement of domestic and foreign private enterprise, the employment of an inflationary development strategy, the recourse to foreign indebtedness on a large scale in the face of growing deficits in the balance of payments, and the shift in the allocation of resources in favour of infrastructural investments in agriculture and communications which distinguished the economic policy of this decade did not, however, alter the basic development pattern of the economy which continued to rely foremost on the encouragement of domestic industry through protection.

The introduction for comprehensive planning in the early 1960s gave a substantial impetus to this industrialisation pattern and initiated a major attempt at extending import-substitution into the intermediate and capital goods categories under successive five-year plans. In the first place, a 15-year perspective plan comprising the first three five-year plans for implementation during the 1963-77 period was drawn, envisaging rapid growth in Gross National Product and employment, internal and external financial stability, and a more equal distribution of interpersonal and interregional incomes (see SPO, 1963:31-52). For the realisation of rapid growth, main emphasis was once again put on industrialisation as the overriding objective of economic policy. This was evident from the macroeconomic targets of both the First Five-Year Plan (1963-67) and the Second Five-Year Plan (1968-72) which, by way of diverting a large portion of total investible resources into industry envisaged rates of growth for this sector considerably above the rates envisaged for other sectors. The First Five-Year Plan, for example, aimed at an annual rate of growth of 12.9 per cent for industry (11.5 per cent for manufacturing) as against 7.0 per cent and 4.2 per cent for GNP and agriculture, respectively (SPO, 1963:121). Although there were also several attempts at export-promotion, import-substitution through protectionist policies was once again emphasised as the main instrument of industrial growth. Whereas protection and import-substitution that followed from it in the 1950s had been very much a by-product of import restrictions imposed in response to difficulties in the balance of payments, import-substitution through protectionist policies became an integral part of the planning process and emerged as a more direct and deliberate instrument of industrialisation policy. Polices towards the private and public sectors during this period, on the other hand, were guided increasingly within the framework of a "mixed economy" based on a more balanced treatment of the two sectors in major fields of economic policy.

Although the pattern of development as broadly outlined above was instrumental in the creation of an expanding industrial sector and thereby in generating a gradual structural change against agriculture, it was also responsible for a number of problems pertaining to the future growth of the economy which made themselves increasingly felt after the mid-1960s. The principal problem confronting the economy has been the familiar one of pursuing import-substitution beyond the limits of available resources and ensuing pressures on the balance of payments. The acceptance of rapid industrialisation as the main objective of economic policy has meant the heavy concentration of resources, some of which imported, on the small base of the manufacturing sector, whilst deliberately depriving the dominant sector, agriculture, of much of its requirements for productive investment, giving rise to a substantial productivity differential between the non-agricultural and agricultural sectors. As the import-substitution process gradually spread from light consumer manufactures into the more demanding area of consumer durables and producers' goods production, however, further growth of manufacturing has become increasingly more dependent on imports of raw materials and capital goods, in view of the inability of the domestic economy to cater for them. The growth of exports from the manufacturing sector, on the other hand, has been hampered by high costs and overall inefficiency as heavy and indiscriminate protection led to the creation of enterprises in both the private and public sectors without due consideration given to their economic suitability vis-a-vis their size, scale and structure as well as the level and nature of domestic and foreign demand for their products. Added to these was the buoyancy of domestic demand in a highly protected market giving way to strong monopolistic and oligopolistic tendencies in the manufacturing sector. The ensuing sluggish export performance of this sector, viewed against the rapid growth of its demand for imports, has therefore, put a heavy strain on the balance of payments and led to a situation in which the bulk of the scarce foreign exchange resources became "frozen" in the very sector which utilised them most wastefully. With exports, as the main source of finance for the import requirements of the manufacturing sector, showing little growth and remaining heavily dependent on the low productivity agricultural sector, the task of alleviating the pressures on the balance of payments created by this inward-looking strategy has, therefore, fallen increasingly on the inflow of foreign resources on a large scale since the late 1940s and the upsurge in emigrant workers' remittances after the early 1960s.

In view of the heavy dependence of future growth on these two "external", sources of finance and the uncertainty surrounding them, it should have been clear by the late 1960s that the basic problems facing the economy were to a large extent the logical consequence of the trade and industrialisation policies followed since the early days of the Republic, especially the selection of a highly protectionist import-substitution strategy discriminating heavily against the agricultural and export sectors. Although there was an urgent need for a reassessment of the inward-looking strategy, particularly policies vis-a-vis the exchange rate, interest rates, intersectoral allocation of resources and terms of trade as well as export and import regimes, this was not called for on grounds of the pace of growth achieved. For the strategy had an impressive record in this respect, responsible for an average annual rate of growth of 5.8 per cent during 1950/52-1968/70 and 6.3 per cent during 1963-70 as well as for the creation in a relatively short-time of a sizeable industrial sector and for appreciable improvements in important fields like transport, communications, energy, and education. [2] The criticism of the inward-looking strategy could therefore be directed not so much at the quantity of growth so far achieved but at its quality and sustainability in the future. As pushing industrialisation beyond the limits of available resources resulted in heavy import dependence of production and new investment and imposed a big balance of payments constraint on the future growth of manufacturing output, the trends towards rising unemployment and the absence of any major sign of the reduction in the sharp inequalities in the distribution of income represented a clear manifestation of the failure to realise main economic policy objectives with strong welfare implications.

3. THE OIL SHOCK AND THE 1978-79 ATTEMPTS AT STABILISATION

Turkey entered the 1970s with a major (66.6 per cent) devaluation of the lira and the signing of the Annexed Protocol setting out the conditions for Turkey's full membership to the European Economic Community by the mid-1966s. As a result of the devaluation as well as good weather conditions in agriculture, there was rapid growth in exports during 1970-74 which increased from $588.5 million in 1970 to $1,532.2 million in 1974. [3] Thanks to the devaluation and the large increase in the net outflow of workers, there was even a sharper increase in workers' remittances which rose from $273 million to $1,426 million during the same period. As a result, the current account balance turned from a deficit of $58 million in 1970 to a surplus of $484 million in 1973. Given the long-term structural problems facing the economy and the expansionary trends and expectations surrounding world trade, one would have expected Turkey to make the necessary adjustments to her trade and industrialisation policies with a view of removing the biases of the inward-looking strategy and devising an export-oriented policy framework. Instead, the planners, encouraged by the drastic improvements in the external payments situation, drew up the Third Five-Year Plan (1973-77) which, by and large, reflected the inward-looking bias of earlier plans, this time with even greater emphasis on import-substitution in intermediate and capital goods to attain a targeted rate of growth of 7.4 per cent per

TABLE 1

Selected Economic Indicators, 1970-1980

Year	GNP (1)	Total Fixed Investment (2)	Exports (3)	Imports (4)	Terms of Trade (5)	Current Account Balance (6)	Short Term Debt (7)	Wholesale Prices (8)	Annual Emigration (9)	Workers' Remittances (10)	Real Wages (11)
1970	125.4	74.5	588.5	947.6	-	-58	-	6.7	129.6	273	-
1971	138.2	71.0	676.6	1170.8	-	-38	-	15.9	88.4	471	39.32
1972	148.5	78.1	885.0	1562.5	-	15	19	18.0	85.2	740	38.03
1973	156.5	91.7	1317.1	2086.2	100.0	484	279	20.5	135.8	1183	41.37
1974	168.0	103.8	1532.2	3777.6	92.0	- 719	216	29.9	20.2	1426	41.89
1975	181.4	128.5	1401.1	4738.6	77.7	-1880	1398	10.1	4.4	1312	43.31
1976	195.3	146.0	1960.2	5128.6	79.8	-2301	3441	15.6	10.6	983	49.70
1977	203.0	156.0	1753.0	5796.3	79.4	-3385	6236	24.1	19.1	982	50.13
1978	209.1	140.4	2282.2	4599.0	73.8	-1418	7319	52.6	18.9	983	43.94
1979	208.3	132.7	2261.4	5069.4	73.7	-1239	3556	63.9	23.6	1694	38.03
1980	206.9	123.9	2910.1	7667.3	72.0 (12)	-2461	2439 (13)	107.2	28.5	2071	28.40

Notes and Sources:
(1) billion TL., 1968 factor prices; TUSIAD (1980a:160) for 1970-72 and SPO (1981b:5) for 1973-80
(2) million TL., 1976 prices; TUSIAD (1980a:168) for 1970-72 and SPO (1981b:44-56) for 1973-80
(3) and (4) million $, TUSIAD (1980a:201) for 1970-72 and SPO (1981B:89,99) for 1973-80
(5) based on $ prices (1973=100); Central Bank of Turkey (1981:131)
(6) million $, IBRD (1980:256) for 1970-73 and SPO (1981b:63-64) for 1973-79
(7) million $, credits with less than 3 year duration; IBRD (1980:269) for 1972-76 and Central Bank (1981:135) for 1977-80
(8) 1963=100, percentage increase over the previous year; TUSIAD (1980a:236) for 1970-72
(9) in thousands, IIBK (1980: Table 16 and 1981:3)
(10) million $, Central Bank, *Annual Report*, various issues and SPO (1981b:71)
(11) TLs, calculated on the basis of data on gross average daily wages for insured workers given in Central Bank (1980:189) for 1971-75 and SSK (1981:16) for 1976-80 as well as on the Istanbul Cost of Living Index (1971=100) given in Ministry of Commerce (1981:7)
(12) arithmetic mean for the first three quarters with indices 85.0, 66.2 and 64.9 in that order
(13) provisional, (Table 4, below)

annum during the plan period. [4]

In the end, the improvements on foreign exchange earnings proved to be temporary, and much more importantly, the first oil shock hit Turkey very hard, resulting in the quadrupling of oil prices. Given the heavy dependence of domestic energy requirements on imported oil [5], this led to a sharp deterioration in the external terms of trade, with the index falling from 100 in 1973 to 92.0 in 1974 and 79.4 in 1977, and was largely responsible for the large increase in the current account of deficit during 1974-77. The fact that the oil shock came too soon after the Third Five-Year Plan was set into motion combined with the inflexibility of the the planning mechanism and the ecstacy created by the Turkish intervention in Cyprus and the subsequent arms embargo on Turkey were the main factors accounting for the reluctance to adjust to this major event with oil products continuing to bo sold at heavily subsidised prices in the domestic market and the targets of the plan kept intact. Turkey, instead, resorted to the buoyant international capital markets and borrowed heavily on a short-term basis (Table 1), sustaining an average annual rate of growth of 7.7 per cent during 1973-76 in the face of a severe international recession. The heavy reliance on short-term borrowing, however, proved costly, as the sharp increase in the external debt burden [6] led to a loss in credit-worthiness and, finally, to a full-scale payments crisis in 1977. It was only then that Turkey felt it necessary to adjust to the new set of conditions initiated by the first oil shock.

With short-term external borrowing opportunities largely exhausted, there were only two major options that the adjustment mechanism could consider. Given the strong bias against exports imposed by the structure of protection, the unfavourable external demand conditions created by the international recession, and the wave of protectionism in industrial countries, however, this option could not be relied on to take Turkey out of the crisis in the short-term. Consequently, the second option of relying on borrowing from international organisations and cutting back on the import bill was taken which saw imports decline from $5,796.3 million in 1977 to $4,599 million in 1978 and $5,069.4 million in 1979 and the current account deficit reduced to $1,418 and $1,239 million in 1978 and 1979, respectively. Although imports accounted for a small portion of GNP (9.0 per cent, on average, during 1977-79 [7]), the dislocations caused by reduced import availability were considerable in view of the high degree of import dependence in manufacturing which, in 1978, was estimated at 21.1 per cent in intermediate goods and 48.6 per cent in capital goods. [8] As widespread import shortages led to crucial supply bottlenecks and to a fall in investment, the rate of growth of GNP fell from 4.0 per cent in 1977 to 3.0 per cent in 1978 and to - 0.3 per cent in 1979. Meanwhile, the rate of inflation, in the face of shortages and rising costs of imported materials and a booming domestic demand, rose continuously from 15.6 per cent in 1976 to a massive 63.9 per cent in 1979 while rising trends in unemployment created further pressures in a labour market already suffering from the sharp decline in labour out-migration after 1973 (Table 1).

The unfavourable external environment which characterised the period after 1973 was largely instrumental in shifting the emphasis of the debate on Turkish growth and development once again to external factors and removing further from the agenda the reassessment of the system of protection with a view of giving the economy an outward-looking orientation. Although the Fourth Five-Year Plan (1979-83) drawn up in the midst of this crisis preserved much of the long-term commitment of

Turkish planning to rapid growth through import-substituting industrialisation and envisaged an average annual rate of growth of 8 per cent for GNP [9], Turkey in both 1978 and 1979 became pre-occupied with short-term stabilisation which addressed itself, in the first place, to controlling inflation and the provision of external assistance to finance oil and other essential imports. As a result, economic policy-making became increasingly vulnerable to and indeed interwoven with external finance prospects with the IMF and other financial media reaction to Turkish letters of intent having a greater impact in this respect than the objectives and mechanism of central planning as the realisation of plan targets itself required right at the outset international finance backing in terms of both debt-rescheduling and provision of fresh facilities.

Against this general background, attempts at stabilisation beginning from the second half of 1977 were directed at two main problems, growing balance of payments difficulties and galloping inflation. To cope with the first of these problems, considerable emphasis was put on foreign exchange stringency, promotion of exports and workers' remittances, and increasing the inflow of external resources through foreign private investment and international borrowing on a long-term basis. Apart from reducing imports through the regular import programmes, efforts aimed at saving foreign exchange consisted of severe restrictions on foreign travel and imports of inessential consumption goods, especially by Turkish migrant workers on their visits or permanent return to Turkey, and limitations imposed on imports on credit. In its efforts to promote exports and workers' remittances [10], the government, apart from relying on measures like tax rebates and foreign exchange retention schemes for exporters [11], attached a higher degree of importance to exchange rate flexibility which saw a number of exchange rate adjustments and two major devaluations which took the dollar parity of the lira from 19.25 in September 1977 to 25.00 in March 1978 and, after a period of exchange during April-June 1979, to 47.10 in June 1979 for all international transactions, except petroleum and fertilizer imports. [12] As official declarations reflected considerable liberalisation of the attitude towards foreign private investment, considerable effort was made to obtain IMF backing to improve Turkey's ability to reach debt-rescheduling agreements and obtain fresh facilities to alleviate her balance of payments difficulties became, at that stage, conditional on reaching an agreement with the IMF, IMF backing was conditional on the government implementing certain policy changes to attain domestic financial stability.

Although controlling inflation ranked high also on the policy priorities of the government, IMF's emphasis for this purpose on further monetary restraint, particularly, through limitations on public sector credits and net domestic assets of the Central Bank, were, until their inclusion in the stand-by agreement signed in July 1979, were unacceptable to the government. The measures implemented to tackle inflation until July 1979, and thereafter, were, instead, confined to heavy reliance on price control through administrative means, increases in the prices of commodities produced by the state economic enterprises to reduce their losses, restrictions on public expenditure to keep budgetary deficits as low as possible, tighter control of hire-purchase transactions, and encouragement of household savings through increases in the rate of interest which, for time deposits over one year, were raised from 9 per cent to 12-20 per cent in April 1978 and 20-24 per cent in May 1979. [13]

Although a comprehensive assessment of stabilisation policies implemented during 1978 and 1979 is beyond the scope of this paper, their failure to reach major policy objectives may be attributed to several factors. [14] Although the government showed some inclination towards greater flexibility in interest and exchange rates, the adjustments made were, on the whole, insufficient to make up for the erosion of real rates through inflation. Consequently, with exports and workers' remittances, by and large, stagnant and the inflow of fresh facilities much lower than expected, shortages of imported goods, including oil, were instrumental in aggravating supply bottlenecks while successive devaluations increased costs and contributed to rising inflationary expectations. Government's insistence on price control by decree, on the other hand, led to the accumulation of stocks of a large number of goods for speculative purposes, and unavoidably, to widespread black markets, including some basic consumption goods. This was accompanied by similar trends in both the foreign exchange and financial markets as the erosion of real exchange and interest rates gave rise to a situation whereby a sizeable proportion of loanable funds and foreign exchange resources was diverted into unorganised markets. [15]

As the social and political conflict took an increasingly violent tone, the government's lack of success in controlling inflation and removing shortages was gradually reducing its popular support. Meanwhile, the most influential sections of the private sector were expressing increasing dissatisfaction with the economic policies of the government and were becoming highly vociferous in advocating "the opening up of the economy" by which they often meant increased state assistance for exporting, the easing up of exchange control regulations, and the encouragement of foreign private investment rather than any systematic reduction of the degree of protection they have come to enjoy in the domestic market. Under this socio-economic setting, the resignation of the government following a severe defeat in by-elections saw the establishment in November 1979 of a new government more sympathetic to the private sector.

4. THE 1980 STABILISATION PROGRAMME: OBJECTIVES AND MAIN POLICIES

In the light of the developments outlined above, it was clear to most observers that the deepening economic crisis required the implementation of more effective policies to deal with inflation and balance of payments difficulties than undertaken during 1978-79. To get out of the impasse in the short term, the new government felt that the most urgent task at hand was to get the full backing of the IMF and with it improve Turkey's credit-worthiness in the eyes of the world financial community as well as the World Bank and the OECD. Given the IMF's emphasis on stabilisation through strict monetary control, the government may have felt that it had little option but to enforce such policies to get the much-needed external resources. The government's conversion to the need of a new approach to solving the economic crisis corresponded also with a more favourable political environment for Turkey as events triggered off by the Soviet intervention in Afghanistan were instrumental in assigning a new strategic importance to her in the eyes of her creditors.

Against this background, the government, in only its third month in office, announced a major stabilisation programme in January 1980. The policies introduced as part of the programme may be viewed in three main phrases: (i) The policies announced on 24-25 January, (ii) measures introduced in July shortly after the signing of a new 3-year stand-by agreement with the IMF, and (iii) policies implemented after the military take-over on 12 September. The importance of these phases is, however, in terms of the chronology of events rather than any significant change in the objectives and nature of policies announced under the programme which showed remarkable continuity in all the three phases. It is, therefore, appropriate to consider economic policies during 1980 as a whole and as a part of the stabilisation programme. For analytical convenience, however, policies implemented during the year, can be viewed in two major categories; those pertaining basically to the balance of payments and others, specifically aimed at reducing the rate of inflation.

Efforts directed primarily at alleviating the balance of payments difficulties, apart from those aimed at obtaining external credits, consisted mainly of the encouragement of exports and workers' remittances. The main vehicle used for this purpose was government's commitment to implement a more flexible exchange rate policy in an effort to "maintain the competitiveness of the economy". As part of this policy, there was in January a major devaluation of the lira which took the dollar parity of the lira from TL47.1 to TL70 while multiple rates, with only a few exceptions, were removed. [16] This was accompanied as in previous devaluations by an easing in the number of items in the liberalised list through a reshuffle of liberalised and quota lists and a reduction of guarantee deposits and stamp duty on imports (Central Bank, 1981:53). The fact that the new rate was somewhat higher than certain black-market rates prevailing before the devaluation [17] may be explained by the government's determination to redirect foreign exchange resources into official channels as well as to allow for the expected rise in the price level following the deregulation of most prices as part of the January package. The government's commitment to a unified flexible exchange rate, as reiterated in the stand-by agreement signed in June, continued throughout the year (Table 2) with a number of mini-devaluations which, together with the January devaluation, amounted to a 89.5 per cent devaluation for the year as a whole.

TABLE 2

The Exchange Rate, January-December 1980
(Official Rates in TL per $)

Date	Exchange Rate	Date	Exchange Rate
24 January	47.10 (1)	12 October	82.70
25 January	70.00	26 October	84.80
2 April	73.70	8 November	87.95
9 June	78.00	10 December	89.25
4 August	80.00		

(1) With the exception of imports of oil and oil products and fertilisers and related raw materials which were subject to the lower rate of TL35 per dollar.
Source: SPO (1981b:107)

Export-promotion measures introduced by the government were aimed mostly at widening the scope and effectiveness of existing schemes. As part of this policy, in January duties on imported inputs of manufactured exports were removed (SPO, 1980:26) which was accompanied by the simplification of export procedures through the abolition of export registration and considerable reduction in licensing requirements. Furthermore, exporters whose projects receive the approval of the newly established Department of Promotion and Implementation set up at the Prime Ministry were entitled to foreign exchange allocations to import goods and raw materials needed by themselves or affiliated industries [18] while commercial bank and Central Bank credits to such exporters were exempted from all banking charges (SPO, 1980a:34-39). Similarly, to boost the inflow of remittances, Turkish workers abroad were granted the right to open convertible deposit accounts with a minimum term of 3 years and use them for residential and other investment while time (saving) deposits opened upon the sale of foreign exchange remitted to Turkey by them would receive rates of interest 10 to 15 points above the official rates (SPO, 1980a:25). Government's attempts to attract foreign currency deposits from this source continued in late December when such deposits with 3-24 months duration were granted rates above the comparable Euromarket rates. While there were renewed attempts at liberalising the attitude towards foreign private investment, especially in fields like tourism and oil exploration (SPO, 1980a:19-22), efforts to increase the supply of external credits emphasised better debt management and minimum reliance on short-term debt. In the June stand-by agreement, for example, Turkey undertook not to increase external debt apart form rescheduling, OECD and Central Bank borrowing and not to expand debt arrears.

Anti-inflation policies introduced as part of the stabilisation programme, on the other hand, were aimed at reducing excess demand through strict monetary and fiscal policies. Government action in this sphere stemmed, in the first place, from two major considerations about policies implemented under the previous government. First, the strict price controls, holding prices well below their market equilibrium levels, were the root cause of excess demand and widespread black markets and shortages. Second, holding prices of goods and services produced by the state economic enterprises (SEEs) at artificially low levels was an important element in aggravating inflationary pressures as the losses of these enterprises were often met through heavy recourse to Central Bank funds. As a first step in tackling these problems, the government in January declared its firm intention to rely on "freely-functioning market forces" in the allocation of resources and announced price decontrol in both the private and, with a few exceptions, public sectors. [19] (SPO, 1980a:23,32). The SEEs were granted autonomy in their expenditures, comprising spheres like investment, employment, and wages. As an integral part of this policy SEE prices were sharply increased in January with the rate of increase ranging from 45 per cent in gasoline, 55 per cent in cement and products of state monopolies (including cigarettes, alcoholic beverages, etc.), and 75 per cent in steel to 100 per cent in coal, 120 per cent in electricity, 300 per cent in paper, and 400 per cent in fertilisers (TUSIAD, 1980b:73). Similarly, the government declared its intention to see that agricultural support prices did not get out of reach of international prices. The government hoped that the TL309.1 billion of extra revenue to be generated by these price increases would ease up the pressures on the budget. [20] An even more important source of revenue was expected to come through proposed changes in tax laws intended to rationalise and widen the revenue base of the tax system, primarily through correcting the erosion of tax structure in the face of rapid inflation and improving tax administration. Although these laws encompassing a wide range from stamp duties to real estate tax on the one hand, and income and corporation taxes on the other, did not come into force before the end of November [21], their inclusion in the economic programme of both governments during 1980 may, to the extent that it reflected the government's determination to get inflation under control, have gone some way towards reducing inflationary expectations.

Measures introduced in the field of monetary policy, on the other hand, were half-hearted during the first half of the year with measures confined largely to the setting up in January of the Money-Credit Committee with the task of co-ordinating decisions and monitoring developments in key monetary variable, and a modest increase in interest rates which, for example, increased the lending rates of commercial banks by 2 points for medium and long-term credits and 5 points for short-term credits (SPO, 1980a:24-25). The major development in this sphere came with the signing of the 3-year stand-by agreement under which the government committed itself to a strict monetary programme, and soon afterwards announced the deregulation of all interest rates with effect from the 1st of July. Under the monetary programme, the government undertook to observe limits on the net domestic assets of the Central Bank, Central Bank lending to the public sector, and to monitor actual developments in close co-operation with the IMF. Parallel to these developments, there was an increase in the rediscount rates of the Central Bank [22] and also in minimum reserve requirements on both sight and time deposits together with a rise in associated penalty rates. Although interest rate deregulation in an oligopolistic banking sector soon led to the agreement of the largest banks on a common interest rate policy, the overall result was a big increase in both lending and borrowing rates. Two factors can be held responsible for the sharp rise in interest rates following deregulation. First, with the average annual rate of inflation running over 100 per cent, rates of around 20 per cent for time deposits with a term of 1 year meant heavily negative real rates. Second, as encouraged by these negative real rates, unregulated financial intermediaries comprising mainly the so-called "bankers" began to provide a considerable challenge to the power of commercial banks from outside the official banking system. [23]

The stabilisation programme, the main elements of which were briefly outlined above,was highly controversial receiving wholehearted support by the private sector and its main organisations while organised labour and the Turkish leftist movements, including social democrats, remained bitterly opposed to it. As government spokesmen claimed that there was no alternative to the programme in tackling the problems confronting the economy, the opponents of the programme drew international comparisons and likened it to those implemented in Latin American countries with repressive political regimes and denounced the IMF's traditionally neoclassical monetarist approach as inappropriate for Turkish conditions. At a time when the controversy was becoming even more widespread and heated as the first implementation results of the programme were taken, the military removed the government from power but immediately declared its allegiance to the stabilisation programme "until formidable obstacles in implementation are met". It can, therefore, be argued that the military regime's strong action against controlling violence, the banning of the activities of the most outspoken and militant section of the trade union movement [24], restrictions imposed on free collective bargaining and wage increases, and the dying out of the public debate on the programme may have gone a considerable distance toward further reducing inflationary expectations.

Although the stabilisation programme was presented as a major break away from previous economic policies by its proponents and received as such by its opponents, any sound judgement on the programme, likely to remain in force in the near future, should consider its long-term as well as short-term implication and draw comparisons vis-a-vis the long-term economic policy objectives and pattern of growth as well as in relation to the policies implemented as part of stabilisation attempts in 1978-79. Although some of the policies implemented as part of the programme, especially those aimed at the promotion of exports, workers' remittances, and foreign private investment, bore a close resemblance to those implemented earlier, incentives provided in each case were considerably higher in 1980. Similarly, the reliance on exchange rate flexibility, which in some ways may also be viewed as a continuation of previous policies, reflected, however, a greater willingness to preserve real rates of exchange in the face of rapid inflation through a system of mini devaluations at shorter intervals. Efforts in the same direction during 1978-79 notwithstanding, government's commitment to a more significant development in this respect, with full deregulation of rates in the second half of the year and the sharp reduction in negative real rates representing a major break away from previous policies. But by far the most distinctive feature of the 1980 programme was the extensive price decontrol introduced in both the private and public sectors. It can safely be argued that the structure of prices prior to decontrol was urgently in need of rationalisation as markets for a large number of commodities were characterised by severe disequilibria with prices, determined by fiat, failing to fulfill their allocative task. Neither was it possible to defend administered prices on distributional grounds as they were accompanied by widespread black markets and shortages. Despite these arguments, however, official claims that price decontrol was the only way out seem highly questionable since they ignore the potentially harmful effects of price decontrol in a manufacturing sector characterised not only with strong monopolistic and oligopolistic tendencies but also with widespread inefficiency in both the private and public sectors. The distinctive features of the 1980 stabilisation programme, may in the light of the foregoing discussion, be summarised as increased reliance on market forces through deregulation of prices and interest rates and greater exchange rate flexibility but stricter monetary control and a highly regulated labour market.

Before we turn to the analysis of the initial impact of the stabilisation programme in the next section, it is appropriate to deal briefly with some of its long-term implications, in particular the extent to which it represents a break away from the traditional strategy based on import-substituting industrialisation through protection. Despite much talk about outward-orientation, there is very little in the programme that represents a significant change in trade orientation [25] with the lira remaining overvalued and the system of protection remaining intact. In contrast, strong emphasis on market forces in the allocation of resources and the growth in the influence of the private sector are likely to have long-term repercussions. Although there were periods, such as the 1920s and 1950s when the private sector gained prominence, these were times it was still flourishing under State encouragement and heavy protection. In view of the fact that the private sector has now emerged as an equal partner in development, certain aspects of the 1980 stabilisation package such as price decontrol, the gradual reduction in the role of central planning, the encouragement of private foreign investment on a wider scale, the reduction of public sector investments, the rapid growth of private financial markets, and the growing influence of international organisations with a private sector bias, if continued in the future, are likely to lead to a fall in the overall size and influence of the public sector. [26]

5. THE INITIAL IMPACT OF THE STABILISATION PROGRAMME

Our efforts to assess the responses of the economy to the stabilisation programme during 1980 were bedevilled by two major factors. First, it may, on conceptual grounds, be argued that such an assessment will fail to allow for the time lags in the full adjustment of the economy to stabilisation polices which may be highly significant in a country like Turkey where the dislocations in major markets immediately before the introduction of these policies were particularly large. Second, not only the quality of Turkish statistics which leaves much to be desired, but also delays in their publication and lack of uniformity between data published by different organisations render the analysis of short-term responses a difficult task. In the light of these difficulties, our results in this section may be taken only as indicative of general trends in the very short-term rather than as a definitive account of actual developments.

a) Gross national product and investment

The downward trend in the rate of growth since 1975 continued in 1980 with real GNP falling by 0.7 per cent as opposed to a fall of 0.2 per cent in the previous year (Table 1). With the growth of population estimated at around 2.3 per cent, there was a 3 per cent fall in per capita incomes. These aggregate data, however, conceal considerable variation in sectoral performance with agriculture and construction registering slight growth as opposed to a big fall in output in mining and manufacturing (Table 3). Even services which have traditionally represented the most buoyant categories have, with the major exceptions of government services and financial institutions registered negative growth.

While the prevalence of generally good weather conditions was instrumental in sustaining the level of agricultural production, factors like the slow increase in the inflow of imported spare parts and raw materials, bottlenecks in the energy sector, particularly electricity, early in the year, and the upsurge in industrial strife until September were probably the main reason for the fall in capacity utilisation and output in the manufacturing sector. [27] Despite overall rise in the cost of industrial credits associated with rising interest rates in a generally tight credit market, it seems that the last quarter of the year saw an increase in industrial production which may, in part, be attributed to the improved expectations of producers following the military take-over and the ensuring ban on strikes as well as the greater availability of imports (Table 6, Col.4 below). The rise in output in the face of a generally falling domestic demand, however, was accompanied towards the end of the year with an increase in the level of stocks, particularly in construction materials, consumer durables like refrigerators, washing machines, and to a lesser extent, automobiles and related

TABLE 3

Sectoral Distribution of GNP, 1979-1980
(Values in TL million at 1968 Factor Prices)

Branch of Activity	1979	1980	Rate of Change %
1. Agriculture	44,518.0	45,055.9	1.2
a. Farming	42,985.2	43,409.4	1.0
b. Forestry	993.6	1,015.4	2.2
c. Fishing	539.2	631.1	17.0
2. Industry	43,428.7	41.323.5	- 4.8
a. Mining	4,678.1	4,189.1	- 10.5
b. Manufacturing	34,881.3	33,140.2	- 5.0
c. Electricity, Gas, Water	3,869.3	3,994.2	3.2
3. Construction	12,785.7	12,885.3	0.8
4. Wholesale and Retail Trade	26,813.3	26,304.3	- 1.9
5. Transport and Communications	18,641.8	18,069.2	- 3.1
6. Financial Institutions	4,929.1	5,017.7	1.8
7. Ownership of Dwellings	9,528.3	9,921.4	4.1
8. Private Professions and Services	9,448.8	9,372.0	- 0.8
9. Total Industries	170,093.7	167,949.3	- 1.3
10. Government Services	19,415.5	20,545.5	5.8
11. Total (9+10)	189,509.2	188,494.8	- 0.5
12. Income from Abroad	2,856.5	2,642.3	- 7.5
13. GNP at Factor Prices (11+12)	192,365.7	191,137.1	- 0.6
14. Subsidies	3,497.6	2,489.9	- 28.8
15. Indirect Taxes	19,475.0	18,225.9	- 6.4
16. GNP at Market Prices	208,343.1	206,873.1	- 0.7

(1) Figures refer to provisional estimates based on monthly data for 12 months.
Source: State Institute of Statistics (1981:17)

spare parts, and even in certain food products (For details, see Is Bank of Turkey, 1981:8-11).

Parallel to the generally downward trend in real output, gross fixed investment in real terms which had fallen by 10 per cent in 1978 and 5.4 per cent in 1979 fell by a further 6.6 per cent in 1980, with all major activities, with the exception of tourism and education, registering negative rates of growth over the previous year, ranging from 13.1 per cent in energy and 9 per cent in residential construction to 4.6 per cent in manufacturing and 2.2 per cent in health facilities. [28]

b) Balance of payments

The developments in the Balance of Payments during the year have reflected the familiar picture of large foreign trade deficit closed mainly through workers' remittances in the current account and external credits in the capital account (Table 4). As can be seen from the Table, foreign trade deficit has risen sharply following a large increase in imports in the face of much slower growth of exports. A significant portion of the rise in the trade deficit was due, however, to a further deterioration in the external terms of trade which, after some improvement in the first quarter of the year, continued to fall in the second and third quarters as the rise in the prices of imports exceeded those of exports by a considerable margin. [29]

TABLE 4

Balance of Payments: 1979-1980 ($ million)

	1979	1980 (1)
CURRENT ACCOUNT	- 1239	- 2461
Foreign Trade	- 2808	- 4290
Imports	- 5069	- 7200 (2)
Exports	2261	2910
Invisible Transactions (net)	1559	1826
Interest Payment (3)	- 546	- 667
Tourism and Travel (net)	179	196
Workers' Remittances	1694	2071
Profit Transfers	- 42	- 47
Payment for Services from Project Credits	- 65	...
Other (net)	339	273
NATO Infrastructure	10	3
CAPITAL ACCOUNT	276	2068
Foreign Debt Repayment (3)	- 485	- 583
Foreign Private Capital	86	34
Imports with Waiver	124	91
Project Credits	421	518
Consortium Credits (3)	677	1888
Other Capital Movements	- 547	120
OVERALL BALANCE	- 963	- 393
SHORT TERM CAPITAL MOVEMENTS	470	217
CHANGE IN RESERVES (4)	- 111	- 606
IMF	3	399
NET ERRORS AND OMISSIONS	601	383

(1) Provisional Estimates
(2) The latest figure published is 7667.3 (see Table 3)
(3) Excluding rescheduling
(4) - = increase
Source: Central Bank of Turkey (1981:130)

In view of the fact that the two preceding years had represented a period of high import stringency in the face of a full-scale payment crisis, the level of "suppressed" demand for imports at the beginning of 1980 was probably very high. Although the sharp increase in import prices meant a much slower increase in the quantity of imports than indicated by the data on import values, the indications are that import transfers were carried out with greater speed, especially in the second half of the year. This was evident, apart from an increased inflow during that period (Table 6, Col.4), from an increase in the share of liberalised list items in total imports (Central Bank, 1981:135) as well as from the sharp reduction in the number of items waiting for transfer which fell from 947 in late January to 468 in early October (SPO, 1980:28). The fact that import transfers could be made with relative ease did not, however, alter the composition of imports which continued to reflect the predominance of intermediate and capital goods with imports of oil and oil products alone accounting for nearly two fifth of the total. [30]

TABLE 5

Major Exports, 1979 and 1980 (quantities in thousand tons, values in $ million)

	Quantity	Value	Quantity	Value
Hazelnut	134.3	353.0	100.9	394.8
Cotton (1)	174.5	231.8	204.0	328.5
Tobacco	69.6	177.0	83.7	233.7
Raisins	75.9	114.8	80.3	130.3
Figs	34.9	41.5	32.6	38.7
Total	489.2	918.1	501.5	1126.0

(1) including lint
Source: SPO (1981b:101-102)

TABLE 6

Exports, Imports, Workers' Remittances, and Gold and Foreign Exchange Reserves, 1979-1980 ($ million)

	Exports		Imports		Workers' Remittances		Gold and Foreign Exchange Reserves (1)	
January	216.2	236.3	279.0	420.3	83.1	83.3	641.2	685.2
February	249.3	244.2	398.6	593.3	75.4	173.3	652.1	884.4
March	196.2	233.6	525.8	370.1	55.7	111.7	672.7	927.3
April	187.4	219.0	386.7	366.5	128.6	114.8	661.0	954.6
May	162.5	196.5	335.9	692.5	639.3	103.2	1067.2	871.2
June	167.2	169.3	486.9	556.9	106.2	155.0	1049.3	788.0
July	146.3	167.1	272.0	619.9	100.8	279.4	1177.7	1036.1
August	162.0	180.0	334.1	584.1	111.4	279.5	1138.9	1209.3
September	168.4	219.8	621.9	1137.2	110.7	207.3	1020.0	1266.2
October	174.7	260.7	436.9	903.7	90.5	205.1	920.8	1308.6
November	212.8	326.5	358.4	516.8	99.5	172.5	853.0	1191.7
December	218.2	457.2	642.4	904.3	93.2	185.9	705.8	1208.7
Total	2264.1	2910.1	5069.4	7667.3	1694.4	2071.1	-	-

(1) Sum of reserves at the Central Bank and other banks.
Source: SPO (1981b:99, 80, 71 and 106)

Exports, on the other hand, indicated slow growth with the 28.7 per cent growth in value terms attained over the previous year attributable mainly to the rise in international prices. Although the rate of growth of mining and manufactured exports reached 44.2 per cent and 33.4 per cent, respectively as opposed to 24.4 per cent for agricultural products, there was little change in the composition of exports with agricultural products accounting for 57.4 per cent of the total while manufactured and mining products represented 36 per cent and 6.6 per cent, respectively (SPO, 1981b:101-102). Further disaggregation of export data indicates that Turkey's major exports, accounting for 38.7 per cent of total export receipts showed little growth, increasing by 22.5 per cent in value terms and only 2.5 per cent in quantity terms, as the increase in cotton and tobacco was not matched by others. [31]

The examination of monthly data on exports, workers' remittances and gold and foreign exchange reserves indicates a marked improvement in the second half of the year (Table 6). In the case of exports, growth in receipts over the previous year during the last three months was 72.4 per cent as against only 12.7 per cent during the first nine months with the rate of growth high in all major categories, especially in mining and manufacturing. This improvement in export performance may be interpreted as a response to government attempts at export promotion, particularly through exchange rate flexibility and simplification of export procedures, as well as an outcome of the fall in domestic demand and the piling up of inventories, and the diversion of trade following hostilities between Iraq and Iran. Similarly, despite the low rate of growth (22.2 per cent) recorded by workers' remittances during the year as a whole, growth in the second half of the year exceeded that in the first by a considerable margin. Although the increase in the number of workers placed abroad may also have had some effect, the main reason for this increase may, apart from the emphasis on exchange rate flexibility, be attributed to the sharp rise in interest rates with effect from July and the interest rate premiums granted as part of the effort to attract migrant workers' deposits into Turkish banks.

With tourism and foreign private investment continuing to make a small impact despite considerable encouragement, the main task of alleviating the balance of payments difficulties have again fallen on debt-rescheduling and the inflow of foreign credits on a large scale which came mainly from the OECD Consortium, IBRD and IMF and to a lesser extent from various Middle Eastern Funds [32] and some socialist countries.

c) Money, credit, and price level

The behaviour of all major monetary variables during 1980 (Table 7) reflected the government's commitment to a restrictive monetary policy aimed foremost at reducing domestic demand and boost household savings. In the face of an average annual rate of inflation of 107.2 per cent, money supply increased by 46.6 per cent during the year while the rate of growth of Central Bank and deposit bank credits was contained at 63.8 per cent and 57.4 per cent, respectively. This was closely in line with the stipulations of the monetary programme for the second half of the year agreed upon in the stand-by agreement in June with the actual increase in the net domestic assets of the Central Bank remaining within the set targets while growth of Central Bank credits to the public sector was actually below the limits envisaged (Table 8). These data, apart from indicating growing monetary restraint, were also highly

TABLE 7

Selected Monetary Indicators, 1979-1980 (TL billion)

	1978 (1)	1979 (1)	1980 (2)
Currency in Circulation	89.5	140.9	216.3
Money Supply M_1 (3)	281.0	439.9	645.1
Credit Stock (4)	513.5	789.5	1202.9
Central Bank Credit	124.9	382.1	626.0
All Bank Credit	354.3	534.9	810.1
Deposit Bank Credit	275.0	431.0	678.3
Total Deposits (5)	279.8	443.4	686.2
of which	191.0	296.3	426.1
Sight (Commercial and Saving)	(68.3)	(66.8)	(62.1)
Sight (Saving)	103.9	143.7	181.5
	(37.1)	(32.4)	(26.5)
Time (Saving)	53.3	91.6	150.7
	(19.0)	(20.7)	(22.0)
Foreign Deposit Banks	4.9	8.1	14.5
	(1.8)	(1.8)	(2.1)

(1) As of 31st December
(2) As of 26th December
(3) Currency in circulation. Total sights deposits and deposits at Central Bank.
(4) Central Bank (direct) + deposit bank + State Investment Bank
(5) Figures in brackets indicate the share of various deposit categories in total deposits.
Source: Central Bank of Turkey (1981b)

TABLE 8

Net Domestic Assets and Public Sector Credits of the Central Bank, 1980 (TL billion)

	Net Domestic Assets of the Central Bank			Net Central Bank Credit to the Public Sector		
		Modified			Modified	
	Limit	Limit	Actual	Limit	Limit	Actual
June-August (1)	548.0	...	524.5	355.0	...	353.6
August (2)	568.0	...	549.4	376.0	...	370.6
September						
December (1)	631.0	641.0	604.1	395.0	418.0	415.2
December (2)	666.0	690.0	608.8	405.0	429.0	421.5

(1) Average of last reporting date in each month
(2) Average of each weekly reporting date during the month
Source: IMF as cited in ANKA Review, 1981

illustrative of the increasing role of the IMF in the domestic policy-making process in both drawing up guidelines and monitoring implementation. Monthly data on currency in circulation and money supply confirm the emphasis on restrictive monetary policy during the year as a whole but, particularly towards the end of the year which saw the restraint on Central Bank credits accompanied by an overall fall in the share of the public sector in these credits (Table 9).

The deficit in the consolidated budget, on the other hand, reached much higher proportions than envisaged with the overall deficit for the fiscal year as a whole estimated to have reached around TL200 billion (IMF, 1981:5) which reflected the severe underestimation of the pace of inflation for the year as well as the failure to put into force the tax proposals designed to increase budgetary revenues. The government's determination to stick to the monetary programme was, however, so strong that confronted with a sharp rise in budgetary expenditures, increasing public sector arrears in debt payment rather than having recourse to Central Bank funds was relied upon.

The above developments were accompanied by an increase in the volume of deposits which was distinctively more rapid during the second half of the year (Table 9) when there was a sizeable increase in interest rates following deregulation in July. It is doubtful, however, whether the increase in deposits represented a net overall growth in household savings as a significant portion of this increase was probably due to a rediversion of funds into official channels following the issue of certificates of deposit banks after July. [33] Although the prevalence of negative rates of interest may account for the failure to generate a significant shift in the composition of deposits in favour of time deposits (Table 9), the rapid increase in the issue of bonds by the private sector during the year [34] reflected, apart from the growing needs of domestic industry for external finance, the responsiveness of households to assets with comparatively higher rates of return.

Despite the reliance on a generally tight monetary policy, the rate of increase in prices during the year as a whole exceeded the rate experienced in the previous year by a big margin (Table 1) which may be attributed to several factors. First, there were heavy pressures on the cost side, which by way of adding fuel to the cumulative effects of inflation in previous years, were probably instrumental in aggravating the inflationary expectations of producers. It seems that successive devaluations in the face of a highly inelastic import demand structure [35] as well as the sharp increase in January in the prices of certain basic intermediate goods that loom large in total manufacturing production were the main elements reinforcing these inflationary trends. Added to these factors on the cost side were factors like the gradual rise in the cost of credit during the year and the fall, by all accounts, in the degree of capacity utilisation in the first half of the year, as aggravated by import shortages and industrial strife until September. The second factor that can be held responsible for the rapid inflation during the year is price decontrol itself which during the short period after its announcement led to price increases of a large magnitude. Although a part of these increases could be taken as a reflection of "suppressed" inflation during the period before decontrol, a sizeable portion may have been due to strong monopolistic and oligopolistic tendencies in manufacturing which may have introduced an important exogenous bias towards price increases through higher mark-ups.

The change in the overall index of prices for the year as a whole, however, conceals, as in the case of workers' remittances and exports, significant variation during the year. As indicated by monthly data (Table 9), the rate of growth of the wholesale price index, after rising

TABLE 9

Money, Credit, Deposits, and the Price Level, 1979-1980

	Currency in Circulation		Money Supply(M_1)		Central Bank Credits		Total Deposits		Wholesale Prices	
	1979(1)	1980	1979	1980	1979(3)	1980	1979(4)	1980)	1979(5)	1980
January	96.5	157.7	284.4	435.1	241.5 (72.0)	393.4 (67.4)	274.3 (20.3)	430.3 (21.8)	4.7	9.2
February	96.0	158.8	290.4	480.6	249.8 (73.2)	421.6 (64.5)	282.7 (20.2)	445.8 (21.1)	4.8	29.3
March	93.0	130.6	297.0	445.0	267.8 (71.6)	431.1 (63.7)	296.7 (19.4)	475.9 (20.1)	5.1	4.4
April	106.0	167.3	317.1	480.1	271.2 (71.0)	464.5 (62.3)	309.0 (18.7)	454.0 (21.4)	7.8	3.5
May	115.4	162.8	337.2	491.5	257.6 (75.8)	460.4 (63.4)	328.7 (22.0)	474.8 (20.4)	4.7	2.9
June	118.4	158.3	353.4	507.3	285.3 (72.1)	486.0 (63.7)	347.4 (22.3)	498.2 (19.7)	7.2	2.8
July	117.7	189.5	360.2	547.5	289.3 (74.2)	517.1 (65.8)	363.6 (21.7)	546.7 (20.3)	3.3	0.2
August	128.1	204.5	360.6	583.2	303.8 (73.3)	541.5 (65.2)	352.0 (22.6)	568.9 (20.5)	3.3	1.3
September	132.3	213.9	377.9	612.7	316.9 (74.7)	567.0 (66.9)	366.5 (21.8)	619.5 (19.8)	3.0	3.5
October	169.0	233.4	430.2	644.5	358.9 (70.0)	604.5 (64.5)	379.7 (21.3)	643.6 (20.0)	5.1	7.1
November	145.4	234.5	414.3	653.0	351.4 (70.0)	618.1 (64.8)	396.5 (20.7)	657.7 (21.0)	7.9	3.8
December	140.9	229.5	439.9	660.0	382.1 (67.9)	626.0 (63.7)	443.4 (20.7)	675.9 (21.3)	4.2	3.1

Notes and Sources:
(1) and (2) billion TL, provisional Central Bank data SPO (1981b:139) and Central Bank (1979:33).
(3) billion TL, SPO (1981b:150-51) Figures in brackets indicate the percentage share of public sector credits in total Central Bank credits.
(4) billion TL, SPO (1981b:146-47) Figures in brackets indicate the percentage share of time (saving) deposits in total deposits.
(5) Percentage change over previous months; Ministry of Commerce (1981:2).

sharply in February, the first month after decontrol, fell almost continuously thereafter until September. The acceleration in the rate during September and October, which may be attributed mainly to the combined impact of the large increase in agricultural support prices and money supply during July-September and the initial uncertainty created by the military take-over in Turkey and hostilities between Iran and Iraq, was followed by a return to relative stability during the last two months of the year.

Although the generally good harvest in agriculture and the increased availability of imports during the second half of the year have also played an important part, the overall deceleration in the pace of inflation may be largely attributed to monetary restraint cutting back domestic demand which was to some extent confirmed also by the relatively slower growth of the consumer price index. [36] It may, in the light of the foregoing discussion and the continued downward trend in prices during the first quarter of 1981, be safely argued that the stabilisation programme has, by reversing the inflationary trends of the past few years, reached one of its chief objectives.

d) Employment, wages, and the distribution of income

The labour market in Turkey has presented the familiar case of the pace of employment creation falling drastically short of the increase on labour supply in the face of rapid population growth, averaging around 2.5 per cent per annum during 1950-80. [37] Although population growth was at the centre of the pressures on the labour market, there has been considerable reluctance in introducing an effective government sponsored population control programme. As a result, the increasing trends toward capital intensity, as encouraged by factor price distortions introduced by the domestic trade and industrialisation policies, together with the rapid pace of urbanisation have led to widespread unemployment and underemployment in both the agricultural and non-agricultural sectors. The severity of unemployment in urban areas would have been far greater had it not been for the absorption of a sizeable portion of labour supply by the small enterprises making up the "unorganised" sector and the services sub-sector, and the migration of a large number of workers mostly to West European countries after the early 1960s. The sharp fall in the outflow of workers through the latter channel, however, has increased the pressures on the labour market which, in the absence of any major effort aimed at manpower planning, was characterised by rising unemployment also among the young and educated members of the labour force. Parallel to these trends in the labour market, there were severe distributional imbalances between urban and rural areas and among different regions with the size distribution of income revealing that the fifth of the population with highest incomes received 56.5 per cent of total incomes while the lowest fifth received only 3.5 per cent (SPO, 1976:1-4).

Although employment, wages, and the distribution of income represent areas in which the availability and quality of statistical information are most deficient, some tentative conclusions may be drawn about the effect of the stabilisation programme in these spheres. First, the poor growth performance of the economy during the year, especially in manufacturing, has by all accounts led to an increase, led to an increase in the level of unemployment which, according to some estimates, rose by half a million and reached around 20 per cent of the labour force (ANKA, 1981). Although the existence of a large pool of surplus labour is widely recognised, this is not borne out by official statistical data on unemployment. According to figures given by the Turkish Employment Service, for example, the total number of unemployed as of the end of 1980 was only 263.400, representing less than 2 per cent of the total labour force. [38] These unduly low employment figures may be explained by the overall slackness of the labour market characterised by the imperfect flow of information and highly limited employment opportunities, which together effectively discourage potential entrants into labour force from even registering as unemployed. Notwithstanding its underlying shortcomings, the data from this source confirm the trends towards rising unemployment during 1980, indicating a fall in employment opportunities and placements by 19.4 per cent and 22.7 per cent respectively as opposed to a 50.2 per cent increase in the level of "open" unemployment [39] despite a 20.6 per cent increase in the number of workers placed abroad with the bulk of unemployed persons (76.4 per cent) belonging to the 15-29 age group (IIBK, 1981:3). Second, there are strong indications that the policies conducted under the stabilisation programme aggravated the erosion of real wages experienced since 1977 which saw them nearly halved during 1977-80 (Table 1). Partly as a reaction to this phenomenon, industrial strife during the period before the military take-over in September was at its peak which saw strikes reach unprecedent heights with days lost through strikes during January-

TABLE 10

Average Daily Wages (1) and Prices, 1979-1980

	1979 TL	1980 TL	Increase %	Annual Increase (2) in Cost of Living Index %
January	210.43	310.80	47.7	96.2
February	210.15	309.33	47.2	117.4
March	245.27	379.27	54.6	120.2
April	244.06	381.54	56.3	117.4
May	272.14	379.33	39.4	113.0
June	270.28	389.22	44.0	101.4
July	281.65	416.36	47.8	89.2
August	293.10	437.94	49.4	88.2
September	292.74	n.a.	--	86.7
October	299.64	n.a.	--	82.7
November	295.03	n.a.	--	76.2
December	307.05	n.a.	--	75.1
AVERAGE	294.31	426.96	45.1	94.3

(1) Insured workers only
(2) Based on the Istanbul cost of living index, 1963=100
Source: Ministry of Finance (1980:139) for 1979 and January-June 1980, and SPO (mimeo) for July-August 1980 for monthly wages, SSK (1981:16) for average annual wages, Ministry of Commerce (1981) for prices.

August 1980 representing more than a three-fold increase over the whole of 1979 [40] (Ministry of Finance, 1980:151). The data available only for the January-August period, however, indicate that increased militancy by organised labour did not prevent the fall in real wages by a large margin as the increase in average daily wages (for insured workers) was substantially below the rate of increase in the cost of living index (Table 10).

The above trends towards declining real wages was reflected also by the fact that the minimum wage which was fixed at TL180 per day in May 1979 was kept at that level throughout 1980. [41] Third, it may, in the light of the developments in employment and wages, be argued that there was further deterioration in the distribution of incomes during the year. Although the rise in agricultural support prices announced by the government in expectation of early general elections may have alleviated the burden of inflation on the agricultural population, the sharp fall in real wages in the face of restrictions on free collective bargaining and price decontrol is likely to have altered the functional distribution of income against labour. Similarly, price decontrol and the considerable leverage that the private sector obtained under the programme, particularly vis-a-vis Central Bank credits and investment and export promotion incentives, may have led to a distributional shift in favour of the private sector. Within the private sector, however, it seems that the bias against relatively small enterprises in the provision of government economic services and other incentives has continued, particularly in the allocation of credit and foreign exchange. Although the rise in costs, especially after the deregulation of interest rates, tended to have an adverse effect on all enterprises, it is likely that the overall effect on larger enterprises was smaller not only because their considerable monopoly power enabled them to pass on the rise in their costs with greater ease, but also because big manufacturing interests were closely interwoven with banking interests with some of the big commercial banks actually owned by them. The latter phenomenon was important also because banking, characterised by big margins between lending and borrowing rates, was perhaps the most profitable and buoyant activity which, together with the mushrooming of other smaller private financial intermediaries during the year, may be taken as a further indication of the growing privatisation of the economy.

6. CONCLUSION

The whole array of economic policies introduced as part of the 1980 stabilisation programme, in its major objective as well as in its impact, bears a close resemblance to the experience of certain other less developed countries. As elsewhere, the avowed objective of these policies was to cope through strict monetary control with galloping inflation and growing pressures in the balance of payments, identified as the two main problems confronting the economy. As an integral part of this diagnosis, restrictive money and credit policies, price decontrol on a large scale were relied upon to bring about a significant reduction in domestic demand which was hoped, in turn, to increase exportable surpluses as well as reduce inflation.

The first results for 1980 indicate that the programme had considerable success in wiping out black markets and shortages, in reversing the inflationary trends, and, to a lesser extent, in initiating an upward trend in workers' remittances and exports, particularly during the last quarter of the year. Although the emphasis on exchange and interest rate flexibility as well as overall monetary restraint have also played a major role, the relative success of the programme in these spheres must be seen in the light of two exogenous factors; the wide acclaim that the programme received in Western financial circles and the subsequent inflow of external credits on a large scale, facilitating an increased import flow, on the one hand, and the military take-over in September reducing inflationary expectations, not least through restrictions on trade union activity and emphasis on wage restraint. Furthermore, the programme had a much less impressive record in other spheres with the policies implemented under the programme, by all accounts, instrumental, not only in increasing distributional imbalances at all levels but also in aggravating the recessionary trends in the economy through a fall in aggregate income and investment and a rise in unemployment despite the increased availability of imports. It was this record, together with growing dependence on external sources of finance and greater emphasis on the private sector with increased reliance on the market mechanism which seemed most in conflict with long-term objectives like rapid growth and industrialisation within a planned mixed economy framework with minimum reliance on external finance but increased importance attached to employment and distributional issues.

Although the short period of time that has elapsed since the adoption of the programme may have prevented the full adjustment of the economy, there are several factors at work which do not augur well for the programme's success in the future. The increased inflow of imports during the year which was perhaps the most important single factor behind the partial success of the programme may prove to be the chief bottleneck in the years ahead as the economy's capacity to import is likely to highly restricted. Given the traditional volatility of the willingness of Western financial circles to provide credit to Turkey, one cannot rely on a steady inflow of resources from this source even in the immediate future, the 3-year stand-by agreement with the IMF notwithstanding. Neither can foreign private investment be relied on in this respect given its traditional reluctance to invest in Turkey which continued in 1980 despite considerable encouragement by the government. Similarly, the growth prospects of Mediterranean tourism in general do not seem promising. [42] Slow growth prospects for world trade as well as both recessionary and protectionist trends in the OECD countries are likely to have an adverse impact on exports and workers' remittances with the outflow of workers to these countries having already come to a virtual halt. These factors, taken in conjunction with the heavy external debt burden on Turkey and the uncertainties regarding the price of oil and other imports, may lead to further difficulties in the balance of payments in the coming years which may, apart from increasing the constraints on growth and employment, start a new period of galloping inflation. The latter possibility may be reinforced by a wave of higher wage claims by organised labour which, after a period of falling real wages, will be hard pressed to recover lost ground with the envisaged liberalisation of political life in the near future.

If the stabilisation programme has yielded poor results in its first year of implementation and is likely to constrained by certain factors in the years ahead, and more significantly, is in sharp conflict with long-term development objectives, there is an urgent need to develop alternative policy packages. Although the development of such a package is beyond the scope of this paper, certain tentative conclusions can be drawn in the light of our present analysis. First, in an effort to better synchronise long-term development policy objectives with short-term stabilisation attempts, there is a pressing need to reactivate the comprehensive planning mechanism. Such a reactivation, which would draw upon considerable experience in the planning field, should lead to a new attitude in planning involving greater flexibility to allow for quick adjustment to changing internal and external circumstances and increased emphasis on project evaluation, and should be equipped with a vastly improved data base. The primary task of the revived planning mechanism should be directed in the short and medium-term at the reorganisation of the SEEs to increase their efficiency on the one hand and at increasing public investment in infrastructure, particularly in the energy and transport sectors with a view primarily of reducing the economy's dependence on imported oil. Although it is likely to be constrained by an unfavourable external environment, an export drive should be initiated with additional incentives granted to sectors and commodities with the highest export potential. Most importantly, the prevalence of unfavourable external circumstances which was the major factor behind the adoption of an inward-looking strategy, should not be used as a pretext for its prolongation. Instead, necessary adjustments in domestic economic policies should be made gradually with a view of giving the economy an outward-looking orientation which, given the heavy dependence of domestic production and new investment on imports, is a pre-requisite for the economy to reach self-sustained growth in the long-run. To this end, the revived planning mechanism, assumed to operate within the existing socio-political setting, should proceed along a broad front, encompassing policy shifts in three major areas. [43].

First, in the absence of additional land that that can be brought under cultivation, future growth in agriculture will have to rely on raising the level of productivity on existing resources which depends crucially on the diversion of sizeable resources to this sector for technical change as well as for institutional (marketing, credit, land tenure) and infrastructural (transport, irrigation, education, technical personnel) improvements. As an integral part of this policy package, there should be a reassessment of the agricultural price support policy which, apart from being a major source of domestic inflation (as the losses incurred by purchasing agencies in the process have come to be financed by Central Bank funds), has, rather like the system of protection in manufacturing, acted as a strong disincentive for reducing costs and adopting new techniques of production. Increased productivity in agriculture would, to the extent that it leads to the growth of this sector's exportable surplus, increase the economy's capacity to import while reducing the level of underemployment, intersectoral migration, and the inequalities in income distribution.

Second, a major effort should be made in reducing import dependence in manufacturing which was, as in many other developing countries, the direct outcome of domestic trade and industrialisation policies favouring the production of capital-intensive and final-stage consumption goods. Apart from the immediate steps suggested above vis-a-vis the energy sector, an important step towards reducing import dependence would be the reassessment of the role of manufacturing branches associated with the production of consumer durables which are generally characterised by heavy import dependence, particularly in "assembly" industries. Although any fall of production in these branches may have welfare implications in the short-term, especially vis-a-vis the level of employment, these considerations are outweighed by the fact that under the existing economic policies any shortfall in imports causes wider dislocations throughout the manufacturing sector and initiates an inflationary process.

Third, more emphasis should be placed on increasing exports from the manufacturing sector. In view of the apparent sensitivity of exports from this sector, along with other minor exports, to domestic inflation and exchange rate policy, a great deal of importance should be attached to preventing big increases in the domestic price level and avoiding overvalued exchange rates. But much more fundamentally, there should, in view of the adverse effects of shielding enterprises from external and domestic competition over long periods, be a major re-examination of the whole system of protection which may entail a gradual planned liberalisation of foreign trade to reduce the bias against exports and encourage industry to break away from its basically inward-looking orientation. This should be supplemented in the selection of new projects by greater importance attached to the ability of enterprises to attain world market competitiveness.

The gradual implementation of the policy package presented above, within the reactivated planning mechanism, although likely to meet with strong resistance from vested interest groups, is essential for the removal of the structural obstacles preventing the attainment of self-sustained growth. Without success in these spheres, relying primarily on monetary control as under the current stabilisation programme, is likely to fail in bringing about continued price level stability and external balance and would instead cause further divergence from long-term development policy objectives.

NOTES

1. This section draws heavily on Senses, 1979, pp.14-20

2. Figures based on data for GNP at factor cost (1968) prices as given in SIS, 1973, pp.44-45.

3. All figures in this section, unless otherwise stated, are based on Table 1.

4. See Ebiri, 1980, pp.222-26, and SPO, 1973 for further details on this point.

5. In 1977, for example, imports of crude oil alone accounted for 65.7 per cent of total merchandise exports. Figures based on SPO, 1979a, pp.27-28.

6. Total short-term debt, comprising credits with less than 3 year duration, rose (in $ million) from 216 in 1974 to 6236 in 1977 and 7319 in 1978, (Table 1, Col.7).

7. Figures based on GNP data in current prices given in SPO, 1981b, p.4, and import data in TL given in TUSIAD, 1980a, p.187.

8. IBRD, 1980, p.263. Figures refer to the ratio of imports to domestically produced goods in total domestic use.

9. The target rates of sectoral growth were 5.3 per cent for agriculture, 9.9 per cent for industry and 8.5 per cent for services. Within

manufacturing, the emphasis was again on intermediate and capital goods which were envisaged to grow by 12.6 per cent and 15.2 per cent, respectively as against only 8.4 per cent for consumption goods. SPO, 1979b, pp.205-206.

10. After May 1979, saving accounts opened by migrant workers were granted preferential rates of interest which, for deposits with a term between 1 to 3 years, were 10 points above the basic rate.

11. In June 1978, the scope of tax rebates was widened which also saw the maximum rate increased from 25 per cent to 35 per cent, while exporters of manufactured goods were allowed to retain 25 per cent of their foreign exchange earnings to finance their import requirements with the rate raised to 50 per cent in April, 1979.

12. Figures are as given in TUSIAD, 1980a, pp.27-34, and TCMB, 1980, p.56.

13. Figures are as given in Ministry of Finance, 1980, p.70.

14. See Olgun, 1978 and 1979, and TUSIAD, 1980a, pp.13-26, on the content and main implementation results of stabilisation policies during this period.

15. As widely reported in popular press, illegal dealings in foreign exchange, as fed mostly by the savings of emigrant workers, under-invoiced exports, and illicit trading in gold, reached very high proportions and created the image of a highly organised capital market with the "market" rate fluctuating several times during the day on the basis of information received from different dealers in town.

16. The only exceptions were Turkey's major agricultural exports and imports of fertilisers and related raw materials, pesticides, and insecticides for which lower effective rate of TL55 per dollar was applicable. Most significantly, however, imports of oil and oil products which enjoyed a lower rate of TL35 per dollar prior to the devaluation were made subject to the new rare of TL70 per dollar. For details, see Central Bank, 1981, pp.54-55.

17. According to SPO, 1980b, p.18, the black market rate prior to devaluation reached a level "as high as TL60-65" per dollar.

18. It was stipulatéd further that such allocations should not exceed the net FOB value of exports pledged by the exporter. SPO, 1980a, p.38.

19. Coal, fertilisers, electricity for the production of aluminium and ferro-chromium only, and services of State Railways and Maritime Lines, declared as basic commodities, were the only items that remained subject to government price control.

20. SPO, 1980b, pp.22-23. Subsequent increases in SEE prices throughout the year have significantly increased this contribution which, by the end of October had reached TL443.1 billion, SPO, 1980b, p.23.

21. For further details on tax laws enacted during November and December. SPO, 1981a, pp.9-114.

22. The rediscount rate (for short-term credits) which had been raised to 14 per cent in February was raised further to 26 per cent in July, Ministry of Finance, 1980, pp.71-73. As an extension of these policies, the banks were allowed to issue certificates of deposits and pay interest on a monthly basis. SPO, 1980a, pp.84-85.

23. Although bankers, whose main activity consisted of dealing in private sector bonds, in some cases, actually carrying out ordinary banking functions like accepting deposits, could be considered as part of Turkey's hybrid capital market, the lack of information on the extent and nature of their activities and the absence of rules and regulations guiding their operations left much to be desired.

24. According to figures given in SPO, 1980b, p.158, out of the 189 workplaces on strike as of 5 September 1980, 162 involved DISK, the trade union whose activities were banned a week later.

25. It is early days yet to say whether the abolition of quotas in January 1981, the renewed interest in joining the European Economic Community, repeated attacks on the exchange control system by the most influential section of the private sector signal such a change.

26. The shift in balance in favour of the private sector was evident also from its proposed entry into fields such as cigarette-manufacturing, hitherto monopolised public sector as well as from renewed demands that the ownership of the SEEs should be transferred to the private sector.

27. Within manufacturing, the highest rate of growth was attained by the chemicals-petroleum and coal-rubber products subgroup followed by basic metals and non-metallic minerals while production in all other branches fell in real terms. SIS, 1981, p.8.

28. Figures are in 1976 prices as given in SPO, 1981b, pp.52-56.

29. According to figures given in Central Bank of Turkey, 1981, p.131, terms of trade (1973=100, in dollars) after rising from 74.8 in the last quarter of 1979 to 85 in the first quarter of 1980 fell to 66.2 and 64.9 in the second and third quarters, respectively.

30. Figure based on import data for the first eleven months as given in Central Bank of Turkey, 1981, p.133.

31. Among minor agricultural products the biggest increase was in live animals and citrus fruit, mostly lemons, while food, chemicals, textiles and transport equipment were, in that order, the manufacturing branches registered the highest growth in export receipts.

32. Figures for 1980 given in Central Bank, 1981, pp.48-49, put the amount of pledged credits by the OECD and the EEC at $1.2 billion, total credits at $1.4 billion, IMF release as part of the 3-year stand-by agreement at $0.4 billion, and debt-rescheduling as part of the July agreement with OECD countries at $2.7 billion.

33. A related phenomenon in this respect was the behaviour of some households which, attracted by the overall increase in interest rates and monthly payments of interest, changed their portfolio from fixed assets with a low rate of return into bank deposits despite the prevalence, albeit to a lesser extent, of negative rates, in an effort to maintain their standard of living during a period of rapid inflation.

34. Bonds issued by the private sector rose from $4.6 billion in 1979 to $18 billion in 1980 as opposed to a fall of 61.1 per cent in the sale of the low-return government bonds, Is Bank, p.31.

35. Although difficult to quantify its full extent, this effect was probably very strong particularly in view of the heavy dependence of the economy on oil whose price was increased several times during the year.

36. The monthly change in the Istanbul Consumer Price Index in the second half of the year was, for example, lower than the change in the wholesale price index in every month except July. SPO, 1981b, p.259 and Table 9.

37. Ministry of Finance, 1980, p.143. Preliminary estimates show, however, that during 1975-80 the rate has fallen to around 2.3 per cent.

38. IBK, 1981, p.2, for unemployment and IBRD, 1980, p.241, for the size of labour force which refers to its level in 1978.

39. An important factor in the rising trend in unemployment was the slow pace of employment creation in the public sector, particularly in the SEEs which is likely to continue in 1981 with the government determined not to increase SEE employment while filling only half of total vacancies in a number of enterprises. IMF, 1981, p.7.

40. The same period also saw a more than four-fold increase in the number of work days lost through lock-outs. Ministry of Finance, 1980, p.150.

41. The minimum wage was finally raised to TL10,000 per month in April 1981. Although changes in the income tax structure announced towards the end of the year went some way towards restoring progression and were tantamount to a rise in net earnings, the fact that the rise in net earnings was below the annual rate of increase in the cost of living index while workers who had already concluded a collective bargaining agreement were excluded from the scope of reduced tax rates highly restricted their impact on real wages.

42. For further details on this point, see IBRD, 1980, pp.xxi.

43. See Senses, 1979, Chapter VIII, for further details.

REFERENCES

ANKA (1981); ANKA Review, Vol.2, Nos.40-58, Ankara News Agency, Ankara

Central Bank (1979); Uc Aylik Bulten (Quarterly Bulletin) II-IV, Central Bank, Ankara

Central Bank (1980); 1979 Yillik Rapor (Annual Report, 1979), Central Bank, Ankara

Central Bank (1981a); Yillik Rapor (Annual Report, 1980), Central Bank, Ankara

Central Bank (1981b); Haftalik Bulten (Weekly Bulletin), 20th April, Central Bank, Ankara

Ebiri, K. (1980); "Turkish Apertura", METU Studies in Development, Nos.3-4, pp.209-253

Foxley, A. and Whitehead, L. (eds.) (1980); "Economic Stabilisation in Latin America: Political Dimensions", World Development, Vol.8, No.11, pp.823-928

IBRD (World Bank) (1980); Turkey, Policies and Prospects for Growth, World Bank, Washington D.C.

IIBK (Is ve Isci Bulma Kurumu) (Turkish Employment Service) (1980); 1979 Istatistik Yilligi (Annual Statistics, 1979), Turkish Employment Service, Ankara

IIBK (1981); Aylik Basin Bulteni (Monthly Press Bulletin), No.50, February, Turkish Employment Service, Ankara

IMF (1981); "Turkey: Review of Stand-by Arrangement", International Monetary Fund, 13th January, as reproduced in ANKA Review, Vol.2, No.58, March 17, 1981

Is Bank of Turkey (1980); Is Bulteni (Review of Economic Conditions), Is Bank, November-December, Ankara

Ministry of Commerce (1981); Fiyat Endeksleri (Prices Indices), Ministry of Commerce, Ankara, mimeo

Ministry of Finance (1980); 1980 Yillik Ekonomik Rapor (Annual Economic Report, 1980), Ministry of Finance, Ankara

Olgun, H. (1978); "1978 Yilinda Turk Ekonomisinde Ana Gelismeler, 1978" (Main Trends in the Turkish Economy, 1978), METU Studies in Development, No.20, pp.91-106

Olgun, H. (1979); "1979 Yilinda Turk Ekonomisinde Ana Gelismeler, 1979" (Main Trends in the Turkish Economy, 1979), METU Studies in Development, No.22-23, pp.91-106

Olgun, H. (1981); "Economic Review of '80", Briefing, January 5

Senses, F. (1979); The Effect of Economic Policies on the Pattern of Trade and Development in Turkey, unpublished Ph.D. Thesis, London School of Economics, London

SIS (State Institute of Statistics) (1973); National Income and Expenditure of Turkey, 1948-1972, State Institute of Statistics, Ankara

SIS (1981); Oniki Aylik Verilere Gore Yapilan GSMH Gecici Ikinci Tahmini (Second Provisional GNP Estimate Based on Monthly Data for 12 Months), 31 March, mimeo, State Institute of Statistics, Ankara

SPO (State Planning Organisation) (1963); First Five-Year Development Plan, State Planning Organisation, Ankara

SPO (1973); Yeni Strateji ve Ucuncu Bes Yillik Kalkinma Plani (New Strategy and the Third Five-Year Development Plan), State Planning Organisation, Ankara

SPO (1978); Gelir Dagilimi 1973 (Income Distribution 1973), State Planning Organisation, Ankara

SPO (1979a); 1979 Programi (1979 Annual Programme), State Planning Organisation, Ankara

SPO (1979b); Dorduncu Bes Yillik Kalkinma Plani (Fourth Five-Year Development Plan), State Planning Organisation, Ankara

SPO (1980a); 24 Ocak 1980 Sonrasi Ekonomik Istikrar Tedbirleri-1 (Economic stabilisation Measures on and after 24 January 19801-1) State Planning Organisation, Ankara

SPO (1980B); Ekonomik Istikrar Tedbirleri, Uygulama Neticeleri (Economic stabilisation Measures, Implementation Results), State Planning Organisation, Ankara

SPO (1981); Ekonomik Istikrar Tedbirleri-2 (Economic Stabilisation Measures-2), State Planning Organisation, Ankara

SPO (1981b); Temel Ekonomik Gostergeler 1973-1981 (Main Economic Indicators 1973-1981), State Planning Organisation, Ankara

SSK (Sosyal Sigortalar Kurumu) (Social Insurance Institution) (1981); Istatistik Yilligi 1980 (Annual Statistics 1980), Social Insurance Institution, Ankara

Thorp, R. and Whitehead, L. (ed.) (1980); Inflation and Stabilisation in Latin America, Holmes & Meier, New York

TUSIAD (Turkish Industrialists' and Businessmen's Association) (1980a); The Turkish Economy, TUSIAD, Istanbul

TUSIAD (1980b); 1980 Yilinin Ortalarinda Turk Ekonomisi (Mid-year Outlook on the Turkish Economy), TUSIAD, Istanbul

Ulagay, O. (ed.) (1980); "Ekonomide Baska Cikis Yolu Var mi?" (Is There an Alternative Way-Out for the Economy), Cumhuriyet, April-May

6 Labour migration in the development of Turkish capitalism

CAROLINE RAMAZANOGLU

> "In the history of primitive accumulation, all revolutions are epoch-making that act as levers for the capitalist class in the course of its formation; but this is true above all for those moments when great masses of men are suddenly and forcibly torn from their means of subsistence, and hurled onto the labour-market as free, unprotected and rightless proletarians. The expropriation of the agricultural producer, of the peasant, from the soil is the basis of the whole process. <u>The history of this expropriation assumes different aspects in different countries, and runs through its various phases in different orders of succession and at different historical epochs</u>."
> (Marx, 1976, Vol.1, p.876, my emphasis)

> "It has been a longstanding Marxist cliche to point out that colonial workers are derived from a peasant origin and continue to keep their peasant connections. But the full implications of this have only begun to become clear recently with the development of analysis centering around women's work and the reproduction of labour power."
> (Omvedt, 1980, p.187)

Labour migration calls to mind Fay's comment on <u>verstehen</u>, "this term would be fine except for the gross confusions that surround it." (Fay, 1974:73) In this chapter I examine the causes and consequences of Turkish labour migration in the 1970s, but accounting for any form of labour migration is a much less straightforward task than it might appear to be, and requires some initial consideration of the general problems involved. [2]

LABOUR MIGRATION: GENERAL PROBLEMS OF EXPLANATION

1. Conceptions of migration

Conceptions of labour migration are necessarily surrounded by confusion because the term is common to different theoretical frameworks, ranging from the extremes of modernisation theory, through various dependency formulations, to more recent assertions of labour migration as the outcome of class struggles (Johnson, 1979), or more specifically as a consequence of the changing structure of agrarian relations of production (Peek & Standing, 1982). The main problem in explaining labour migration, whether in the Turkish or any other case, is one of methodology, and thus one of the adequacy of theory. Rural people around the world are leaving the land in their millions, and crowding into towns, shanties, or peri-urban shelters apparently regardless of whether or not there is adequate accommodation, employment or other means of subsistence to be found there. There are also considerable movements of population within rural areas and across national boundaries. There is, therefore, a major causal problem to be resolved since explanation is required on a number of interrelated levels, ranging from the personal decisions of individuals to migrate, to structural changes in the economy which occur in the course of the uneven development of capitalism.

It is highly improbable that a single cause will adequately account for such a complex phenomenon, and mono-causal explanations are widely dismissed as reductionist, but at the same time the rapid retreat from agriculture which can be observed all over the world is very likely to have some common foundation. An adequate account of Turkish labour migration must not only explain the presence of Turks in Western Europe, Arabia and Australia, but also be able to take account of, for example, the reasons why a very young female labour force has been built up in parts of Southeast Asia while East African factory labour has remained predominately male. [3] Both the widespread similarities and the very specific differences between these movements of population and the labour forces (and marginal or surplus labour) which emerge in their wake, need to be explained. We also need to know whether there are any general processes keeping people on the land, such as the opportunity for productive investment commercial peasant farming or the hopelessness of debt bondage, which also need to be explained.

If labour migration is taken to be the movement of people in search of wage labour, or at least of survival in a cash economy, at a particular stage in the development of capitalism, then the phenomenon is both one of individual decisions being made by millions of people around the world, and also one of rapid and radical transformation of pre-capitalist agrarian relations of production. A rapid increase in the movement of labour out of agriculture is likely to indicate a major disruption in the way in which agricultural production is organised and related to the economy as a whole. That is, changes occur in who produces what for whom, and how.

As the quotation from Capital at the head of this chapter suggests, everywhere capitalism develops, people (N.B., Marx used the term "men"; the implications of recognising peasants as men and women and their children will be taken up below) must come off the land in order to provide an industrial labour force, but, Marx also notes, that the way in which this happens is variable. The processes of the dispossession of the peasantry occurred all over Europe but this was not a total process of transformation everywhere; some peasantry survived as small direct producers within the capitalist system, and it was not a uniform experience. Marx comments on the uniqueness of the English experience and the peculiarity of the northern Italian case (Marx, 1976:876). The implication of this argument is that while labour migration is a necessary factor in the development of capitalism, every occurrence of labour migration has its own history and its own consequences. Both the general characteristics, and the historical occurrence and outcomes, need to be adequately accounted for. While we can characterise different forms of labour migration in general, and compare the Turkish experience with experiences elsewhere, we cannot explain Turkish labour migration at this level of generality. Labour migration must arise at a given stage of capitalist development, but it cannot be assumed that Turkish labour migration can be taken for granted as but one more example of a common process; whether this process is conceived as the breakdown of tradition, the underdevelopment of the periphery, or the articulation of modes of production. Rather than ask "what is the cause of labour migration in the Third World?" we would do better to ask "why has so much emphasis in social science been put on the search for general answers to questions that are both general and particular?"

2. General questions and universal answers

The most obvious response seems to be that while we do need general theory and concepts in order to understand historical processes and events, the general and the whole is easier to grasp than the extremely complex and changing interconnections between a whole and its parts. I am not arguing that we should abandon recognition of the common features of labour migration all over the world, but I that explanations of labour migration tend to look either at wholes or parts, but not at both. Marx pointed out that "the complete body is easier to study than its cells" (Marx, 1976:90), and Mao argued that universality is easier to explain because it is more widely recognised; it is the particularity of contradiction which is not understood, and therefore the relationship between the universal and the particular is not grasped (Mao, 1967: 316). [4] Foucault too, although critical of Marxists, draws attention to the importance of the distinction between being able to explain what is happening at a given moment, and the quest for philosophical universals (Foucault, 1982:216). Social science becomes confused when these two activities are not distinguished from each other and when the revelation of philosophical universals become more prestigious than explanations of what is happening.

If the place of labour migration in the contradictory development of capitalism is to be properly understood, much greater flexibility is required in conceptualising the complexities of the relations between individual experience and general process than, for example, the claim that "migration obtains under conditions of lessened social constraints" (Kubat and Hoffman-Nowotny, 1982:222), or that central capitalism mechanically produces reservoirs of cheap marginal or surplus labour (see comment on Meillassoux, O'Laughlin, 1977). While general issues must always be explained, to rely on universal generalisations alone serves to avoid the need to explain what is actually happening. Uncritically used, generalisations like "underdevelopment" or "peripheral capitalism" (or "nature" and "tradition" in other contexts) can become substitutes for explanation.

Accounts of Turkish labour migration have tended not to approach explanation with issues such as these in mind, so that a number of problems about the causes and consequences of Turkish migration remain unresolved.

EXPLANATIONS OF TURKISH LABOUR MIGRATION

While there is a vast and diverse international literature on labour migration, the bulk of this is empirical work, generally based on uncritical acceptance of the implicit or explicit assumptions of modernisation theory. The works of such established international authorities on Turkey and migration, as Daniel Kubat, and Nermin Abadan-Unat illustrate the dominance of this viewpoint in Turkish studies. Explanations of migration drawing on dependency theory or the articulation of modes of production debate have been less in evidence, and less influential. Turkish labour migration has most often been accounted for in terms of the factors internal to a traditional community which cause people to leave their farms or villages; the external factors which entice them into other areas or countries (not always urban), and the factors which cause some individuals rather than others to leave. The weakness of the explanations given by this school are derived from the weaknesses of functionalist theory in general, which are widely discussed elsewhere, and I will only comment on some of the issues which are particularly relevant here.

It is not possible in one article to review all the work on the causes of Turkish migration, let alone the very considerable and polyglot work on the lives of migrants abroad, and the social consequences of migration within Europe. The emphasis in this literature, however, has tended to be on the reasons given by migrants for leaving the rural areas, and on the socio-economic costs and benefits of labour migration (with a shift from an initial optimism to much more pessimistic views of the costs of mass migration both within Turkey and to the labour-importing areas). (See for example, Aker, 1972; Paine, 1974; Miller and Cetin, 1974; Munro, 1974; Abadan-Unat, et.al.; 1976; Karpat, 1976, and the various contributors to Krane, 1975 and Abadan-Unat, 1976)

The problem with these accounts of migration is that a general question on the causes of migration tends to be posed, which requires a general causal process to be discovered. This not only falls into the trap of reductionist logic (usually some form of reduction to a generalised economic condition) but also leads to considerable empirical confusion, since different causal processes are then discovered in different studies. In many Turkish studies the causes of mass migration are thought to be so obvious that they can be briefly summarised without discussion. Interviews with migrants provide insight into motivations and reasons for some rather than others leaving home, and structural causes are then deduced, usually from statistical sources. The following summary is a fairly typical one from the middle of the 1970s:

> "The phenomenon of workers' emigration from Turkey stems from economic necessities. It is a part of the international population movements that are created by economic development. The Turkish economy could no longer bear the burden of the underemployed or unemployed population and has pushed them away into the

> international labour market, starting from 1960. There is a close similarity between this movement and the movement of population from the rural to the urban areas"
> (Keles, 1976, p.169)

Although this view may seem plausible, it incorporates a number of assumptions which need to be examined. In particular, the connection between unproductive population and migration is not explained. High rates of migration may be correlated with local unemployment, but the causal connection is inferred from an implicit acceptance of modernisation theory. There is also the empirical problem that emigrants up to the early 1970s were disproportionately employed, skilled, educated and from the most developed regions of Turkey.

Much empirical research in this tradition consists of establishing correlations and then inferring causes without further justification. The causes of migration are then taken to be such factors as poverty, unemployment, landlessness, overpopulation etc., which do not require investigation. Abadan-Unat, one of the most prolific of Turkish writers on migration, comments on:

> "a migratory wave from newly industrialising countries due to open or concealed unemployment"
> (Abadan-Unat, 1974, p.362)

In a slightly later article, however, she offers another cause, pointing to Turkey's high birth rate, and attributing mass migration to the explosive growth of population (Abadan-Unat, 1976:5).

Turkey's birth rate was indeed high at this period, if compared to those of most European countries, but there is no good reason for drawing conclusions from correlations between population growth and labour migration without further explanation. In 1971, for example, there was a striking similarity between the population figures for Turkey and for Uganda, in terms of birth, death and annual growth rates (Matras, 1973: inside front cover). If the causes of labour migration were primarily demographic we should find a correlation between the similar "explosive" characteristics of population growth in Turkey and Uganda at this time, and their rates of labour emigration. Up to this stage, however, Uganda was a net importer of labour, while the mass exodus of Turkish labour to the cities and abroad was still growing.

To understand the differences between labour migration in Uganda and in Turkey at this time, we have to look not at underemployment or population growth as isolated factors, but at the totality of social, ideological and economic relationships within which migration occurs. This involves identifying the extent of capitalist development in a social formation, and the relation of the social formation in question to the rest of the world system of capitalism. Such analysis would have picked out differences in the transformation of agrarian relations of production, which was occurring very rapidly and unevenly in Turkey, prior to 1971, but only to a limited extent in Uganda.

We might equally well ask why did Yugoslavia with its low birth rate send workers to Western Europe, when Albania with its high birth rate did not? In order to answer this question, one would need to look not at population characteristics, but at the many ideological, political, social and economic differences within Albania and Yugoslavia, and at their positions in relation to the rest of the world. If population growth causes people to abandon agriculture we need to be able to explain why it has this effect, whether it always has this effect, and how variations come about. In the 1970s, dependency theory displaced modernisation as the dominant theory of Third World development, and gave rise to a view of Turkish migration as stemming from a reservoir of surplus labour which had been developed to meet the needs of Western European capital. Although this has become the dominant view elsewhere, and is popular in Turkish development studies, it has been less influential in explanations of Turkish labour migration. Berberoglu, for example in presenting a dependency theory analysis of Turkish capitalism (or rather of the lack of it) does not consider migration as a separate issue (Berberoglu, 1982). In spite of the central role played by migration in the process of proletarianisation, Berberoglu is primarily concerned with the class interests of Turkish workers as against their imperialist exploiters, the international bourgeoisie and its Turkish allies.

Like modernisation theorists, dependency theorists have looked at labour migration as a general phenomenon for which there should be a general explanation. Obviously some general conceptualisation and classification is necessary, otherwise we would not be able to recognise, for example, the movement of Kenyan labour into Uganda, and the emigration of Turks to West Germany as comparable types of labour migration, but dependency theory remains at this general level of conceptualisation, and is unable to take proper account of the specific histories and specific consequences of different movements of labour. This emphasis on the common features of labour migration in different situations, and the search for typical processes in a transition to capitalism has meant that only certain questions have been asked about what is happening. Dependency theory does not ask the critical questions which enable us to explain the differences as well as the common features which characterise these mass movements of population. In one chapter I cannot take up all the issues which need detailed reconsideration, but in the section which follows I examine the implications firstly of taking gender into account, and secondly of looking at the necessity of state intervention in the development of labour migration.

LABOUR MIGRATION IN THE UNEVEN DEVELOPMENT OF CAPITALISM

While I have argued that Turkish labour migration has its own history and consequences, it is also part of a more general process of uneven capitalist development in which social formations which have developed capitalist agriculture and industry relatively recently, share common characteristics. As the second quotation at the head of this chapter indicates, labour migration cannot be understood without recognising that peasants are not all men. Much of the subsistence production which sustains the poor of the world, and has facilitated migration in many ways, is women's work, and many migrants are women. Rather different questions about the nature and consequences of labour migration are raised by treating women as working people rather than as the invisible appendages of men, and I discuss these in the section which follows. A second area of necessary questions concerns what has happened to the related processes of class formation and the practice of state intervention in labour migration, which I discuss subsequently.

1. Labour migration and the sexual division of labour

There is relatively little to be said in general about the difference the inclusion of gender can make to explanations of labour migration. I consider below some of the empirical reports of women's work and the changing sexual division of labour in Turkish villages which raise specific problems of understanding the relations between individual behaviour and structural changes. In general, however, it is necessary to consider what questions need to be asked. The tendency to use the term "men", as Marx did, in place of "people", has had the dramatic effect of rendering women invisible, and reducing their work to "family labour" or "housework", without considering in each case what it is that women do, for whom and why. Kudat has argued that in Turkish agriculture women's labour has matched that of men, but that their productive labour has not been recognised as such (Kudat, 1982:293). Even where labour migrants are all, or predominately, male the phenomenon of migration will leave women working in rural areas. Male emigration from village agriculture can mean impoverishment of the women and dependent children they leave behind or, more probably, it can mean women working as never before to maintain themselves and their children.

> "Peoples' leaving did not diminish what work had to be accomplished but merely left fewer hands to accomplish it"
> (Yenisey, 1976, p.335)

If, on the other hand, women migrate alongside or in place of men, the consequences in wage employment and in rural labour can be dramatic; leading to the creation of women workers who are often disadvantaged in the labour market compared to working class men; to unmarried women competing with male labour; to whole villages being abandoned; to struggles by women within their families to gain autonomy.

There are considerable differences around the world in the sexual division of labour in agriculture. While it is common for a sexual division of labour to exist, the exact tasks which are considered male or female vary considerably (Deere and de Leal, 1982). There are differences within Turkey in exactly what men and women do, and in the rigidity of the division between male and female tasks, but it can be stated that in general peasant women work, and work productively. Any exceptions to this generalisation will need to be accounted for, and not obscured by the unthinking assumption that women are dependent on the work of their men. Women as economically dependent housewives generally appear with the rise of skilled industrial wage labour and the development of a bourgeoisie. Although even here women can be more economically active than might appear (see for example, Kiray, 1981).

Since men and women both work in agriculture, but at different tasks, the impact of migration depends in part on which sex migrates. One cause of migration is the dispossession, impoverishment and differentiation which the development of capitalism brings to the peasantry, and one consequence is pressure on the sexual division of labour. While it is possible for male migrants to take on female tasks, for example when living in workers' dormitories, it is much more common for migration to mean women taking on men's tasks in addition to their own. A common enough consequence of capitalist development.

Bujra has pointed out that where male labour has been proletarianised in Africa, women have not become dependent domestic labourers, but have developed petty commodity production and trade. Women have had to intensify their labour in subsistence production in order to feed their families, and produce surpluses for sale. In the process family life can be threatened or even destroyed (Bujra, 1983).

If women's labour is made visible and the impact of migration on female labour is recognised, it can be seen that there can be little incentive for peasant households to turn working women into dependent housewives. Female dependence is a luxury that few can afford. The pressure will be on women to work harder in the fields, to find paid employment, to go into domestic service, to find out-work or to help run a family shop or small business; many different patterns are likely to be created in response to different situations and cultural pressures.

A peasant household may depend on migrants' remittances to employ labour or hire machinery, or may migrate as a unit to avoid impoverishment. There are critical differences for the development of capitalism between forms of male migration which leave women engaged in subsistence production, and responsible for the maintenance and reproduction of the labour force, and female migration which leads to women abandoning agriculture and moves the reproduction of the labour force to urban areas. The appearance of child scavengers in the cities can indicate a radical breakdown in agrarian structures, leaving women unable to feed their children. Since urban work does not necessarily mean abandoning household production, the processes by which women leave the land are complex. While Hershlag labels the whole development of Turkish agriculture as enigmatic (Hershlag, 1968), Bryceson argues that the way in which women come off the land is particularly enigmatic. Women peasants do not have the same control or possession of the means of production as men so that the effects of changes in relations of production can be different for men and women (Bryceson, 1980). While there are diverse consequences, in general it might be argued that the upheaval in agriculture, of which labour migration is one part, will serve to differentiate the peasantry, increasing poverty, but also possibly generating strata of small capitalist farmers and dividing women by class. Women can appear as capitalist farmers, entrepreneurs or petty-commodity producers during these changes (Bujra, 1983; Starr, 1984). Other women will be reduced to subsistence, starvation, or be driven off the land. Women can enter the proletariat as workers, but usually in the lower occupational strata, and most will have to find employment outside formal wage labour (see e.g. Bujra, 1979). As women become income earners in their own right, and girls receive formal education, relationships within family units will come under pressure, with struggles intensifying within the family for female autonomy (Cliffe, 1979; Bryceson, 1980).

As capitalism develops, the whole notion of the family has to be reconsidered. While the nuclear family is widely thought of the norm for "developed" societies, in practice capitalism tends to vary rather than standardise domestic kinship units. Labour migration can destroy the bases of kinship units, or can create new family forms. Male migration can lead to new demands on extended kinship networks to act as economic units; those who remain in rural areas take over agricultural production and receive remittances from those who migrate. Long term migration with no facilities for urban settlement can virtually destroy existing family units, and permit the development of wage labour without a family wage. It could also be argued that migration creates conditions which will transform ideas on the sexual division of labour and the value of women's work, but whether or not these changes will improve the position of women vis-a-vis men is quite another matter.

As relations of production in agriculture change, there are limits to the range of consequences that can ensue, but in order to grasp what is happening throughout society, rather than just to men, the involvement of women in these changes will have to be accounted for. All that we can expect in general, therefore, is that contradictory processes such as these will be set in motion, and that the outcomes of the particular historical combinations of factors which emerge are likely to be variable, and will need to be investigated. Examination of the sexual division of labour does not, therefore, provide us with general answers, rather it poses questions about the ways in which capitalism develops in different circumstances, which can only be answered through empirical investigation.

2. Proletarianisation and state intervention

State intervention is generally decisive in shaping the direction of capitalist development in the late industrialising social formations, and their peasantries have taken to different forms of submission and resistance. While there are common problems facing states which are pursuing a course of capitalist development, and a limited range of solutions, there are still variations in the transformation of agricultural relations of production giving rise to different patterns of migration. The way in which the state intervened in the creation of surplus labour in Turkey was historically specific, but it can be stated in general that the states of newly industrialising social formations will be drawn into intervention, whether intentionally or not, and thus embroiled in attempts to control the rise of a working class.

> "The fact remains that the early stages of industrialisation in all market economies have been associated with rapid rural-urban migration, which in turn has been linked to the transformation of traditional rural production relations impeding the growth of agricultural output and of urban-industrial employment. This migration, typically accompanied by urban unemployment and poverty, has undoubtedly been stimulated by government policies introduced to accelerate economic growth."
> (Peek and Standing, 1982, p.28)

Peek and Standing arrive at their notion of what is typical by conceptualising the forms of state intervention which have actually occurred in different areas, and the reasons for these interventions, rather than inferring typicality from an uncritical acceptance of a notion of peripheral capitalism. That is, states do typically intervene in the creation of migrant labour forces because the measures to which late-industrialising states can resort in order to stimulate economic growth, are considerably constrained by their disadvantaged position in the world capitalist system. The range of economic strategies which are open to them will necessarily draw labour off the land. While each case will be different, some general level of conceptualising changes in relations of production is necessary, as without these general concepts (what Marx called "sensible abstractions" Marx, 1970:190) what is happening cannot be observed.

One of the main differences between Turkey and the most developed parts of Europe is not that Turkey has simply been underdeveloped by the forces of imperialism, as Berberoglu, for example, suggests, but that Turkish agriculture has been drawn into the world capitalist system in a very different manner from the transformation of agriculture in the first capitalist economies (bearing in mind the qualification that even in early capitalism this was not a standard process). Turkey's small direct producers exist in a social formation dominated by the capitalist mode of production, but it does not follow that they are fully dispossessed, or that they are fully proletarianised. Labour migration should be seen as a key variable in the development of capitalism, rather than as a standard process.

The general process of proletarianisation, of which labour migration is a part, can now be dramatically rapid and inflexible, but since it cannot easily be completed, peasant agriculture persists in different ways and with different consequences. The contradictions which are generated by incomplete proletarianisation are generated within different social formations, presenting their states often with insoluble problems. The development of capitalist relations of production in agriculture will necessarily transform agrarian structures, but not always in the same way. There does not have to be a standard form of "rural exodus which tends to prevail over rotating migration" (Meillassoux, 1981:139). Incomplete proletarianisation is a widespread process, but it does not follow a predictable chronological sequence. It can take different forms, can encompass men or women or both, can generate different ideologies, which are variously related to class practices and consciousness, and is generally very difficult for the state to control.

State intervention is never a simple process since the state is not a harmonious unity. The contradictions generated by the uneven development of capitalism, develop within the state. Standing's work shows very clearly, both in general and in the case of Guyana how contradictory interests within the state can develop over the supply of labour. In Guyana, a conflict developed after the abolition of slavery between different fractions of capital. The estate owners tried to use the state to subordinate the peasants into an immobile labour force who could provide cheap labour by growing their own subsistence as well as working for wages. The rising industrial/commercial bourgeoisie, however, wanted the state to encourage a mobile labour force divorced from the land (Standing, 1980). The conflict between agricultural and industrial interests within the emerging Turkish bourgeoisie are discussed by Huseyin Ramazanoglu in Chapter 2 above.

Elsewhere Standing refers to migration as an "index of a social formation under stress" (Standing, 1979:42) to indicate that labour migration on a large scale appears with the transformation of pre-capitalist agriculture, and that the role of the state in the subsequent upheaval is variable. Labour migration can contribute to the stagnation of agriculture by allowing unproductive peasant agriculture to survive. Omvedt argues, for example, that in India the colonial state played an active role in maintaining pre-capitalist relations of production in agriculture by supporting the dominant position of the landlords (Omvedt, 1980). The state may depend on the peasantry as a tax base, and act to discourage the movement of labour off the land, in spite of the needs of industry for labour and new consumers. Ambivalent land reform policies are indicative of conflicting interests represented within the state. The threat of political unrest in the urban areas may seem a greater threat to stability than destitution in the rural areas, and the state may act accordingly. State policies cannot simply reflect class interests where struggles for dominance between class fractions are fought out within the state.

State intervention in the creation and control of labour migration in Turkey cannot, therefore, be deduced from general theory. In order to account for the phenomenon of mass labour migration in Turkey in the 1970s, and its consequences, we need to turn from general issues to consider the specific contradictions which have developed in Turkey, within an appropriate totality of relationships.

THE SPECIFICITY OF THE SOCIAL FORMATION: TURKISH LABOUR MIGRATION AND THE CRISIS OF TURKISH CAPITALISM, 1970-1980

The development of labour migration whether in Turkey in the 1970s, or elsewhere, cannot be taken for granted as a mechanical or harmonious process. Since there are specific mechanisms which generate and reproduce migrant labour systems, as well as the general processes which are common wherever capitalism develops, we need to turn from looking at the movements of labour migrants solely in terms of experiences and motives of (male) labourers, or at the generalised needs of central capitalism, to look at the totality of the real relationships and processes within which these experiences, motives and needs develop and are constrained. The history of the causes and consequences of labour migration is too complex to account for in full here, so I have focussed on three interrelated processes which, I would argue, are fundamental to an adequate explanation of what happened in Turkey in the 1970s: changes in the social organisation of production, variations in the sexual division of labour, and the contradictory attempts by the state to control the development of a Turkish working class.

1. The transformation of agrarian structures

The forms of labour migration which Turkey experienced during the 1970s arose from the transformation of pre-capitalist relations of production, especially in agriculture, and this process obviously did not start in the 1970s. Turks had been migrating to urban areas for some time, as Keles points out above, and the historical roots of seasonal migration go back to the Ottoman period (Hinderink and Kiray, 1970; Karpat, 1976). There was also considerable movement of population, and thus of labour, during the upheavals of the War of Independence, and the foundation of the Republic.

The economy of the Republic at its inception was primarily agricultural, with its industrial and commercial sectors shattered by war. Agriculture was also affected by the wars, but was still organised primarily into great estates, or peasant subsistence units, with some nomadic populations and some intermediate farming. As industry and commerce developed, a small working class was brought into existence, but the bulk of the working population remained on the land as "feudal" tenants or independent farmers into the 1960s. Paine draws on official statistics to point out that while the proportion of the economically active population employed in agriculture fell steadily during the sixties, by 1972, two thirds were still engaged in agriculture, and only 14 per cent in industry and construction (Paine, 1974:33ff). This agricultural economy was, however, highly diversified in terms of the organisation of production, at village level. Even the term "village" (koy) although widely used, suggests a structural uniformity which did not in fact exist (Tutengil, 1972).

> "Rural as well as urban settlements in Turkey show differences in social structure to such a degree that almost each settlement is a type in itself." (Eberhard, 1954, p.992)

In the aftermath of the War of Independence, agriculture had to be reconstructed within the boundaries of the new republic, which meant settling the nomadic population, repossessing land vacated by Christians (transferred to Greece and the Balkans in exchange for their Moslem peoples) and by the massacred Armenian landlords. The detailed differences in development which these different circumstances created, can be seen in studies which document local changes. Marked differences in the development of capitalism in four Cukurova villages were documented, for example, by Hinderink and Kiray. This was an area which had previously been farmed by nomads compulsorily settled by the Sultan, or immigrants from the Balkans; by large Armenian and Arab landlords, using seasonal migrant labour from the Anatolian mountains, and sharecroppers; or estates owned by the Sultan. The penetration of capitalism into agriculture after 1923 led to increased differentiation between the villages rather than to common experiences of development (Hinderink and Kiray, 1970).

Capitalist agriculture, also developed very unevenly in the rest of Turkey, with marked differences between, for example, the intensive market gardening of Western Turkey, low productivity peasant production, the emergence of small, mechanised capitalist farms, the development of state farms, and the survival of some "feudal" estates. By the 1950s, there were marked differences in levels of investment and mechanisation; in the marketing of crops, (e.g. whether for domestic or export markets) and in the use of wage labour, "feudal" labour, sharecropping and family labour. Nevertheless extensive mechanisation did take place which had the dual effect of intensifying the differentiation between mechanised and non-mechanised agriculture, and also of displacing labour as agriculture became increasingly capital-intensive (Karpat, 1960; Stirling, 1965; Hirsch, 1970; Hinderink and Kiray, 1970). Ulusan points out that government attempts to reduce the inequalities between agricultural and non-agricultural sectors were largely ineffective. By 1973, 22 per cent of households were landless, and of those with land, more than 80 per cent of farms were too small to generate the officially defined subsistence income. (Ulusan, 1980) Paine estimated that two million people out of an economically active population of 16 million were unemployed by 1973 (Paine, 1974:36).

The development of wage labour in capital-intensive industry remained limited, and immigrants to towns were more easily absorbed into service industry, petty commodity production and trade, than into the small, industrial labour force. The towns provided a range of income-earning opportunities both for the impoverished peasants of the centre and east, and for those who had easy access to the cities. The development of communications, and the presence of pioneering migrants already in urban areas, encouraged the rapid growth of shanty towns. These <u>gecekondu</u> (shanty towns varying from self-built shacks to substantial properties which eventually received mains services) grew to the extent of providing whole villages with an instant alternative to agrarian poverty. During the 1960s small direct producers began to leave the land in accelerating numbers, although with complex local variations in migration patterns. Most of these moved to Turkish towns and cities, but a growing number began to take contracts abroad as the post-war boom in Europe brought labour recruiters to Turkey. Once a core of migrants had been to Europe, known paths of migration were opened to those who came later.

By the early 1970s, the development of Turkish capitalism had brought considerable investment in profit-making agriculture in some parts of Turkey, accelerating the differentiation of the rural population, and the transformation of dispossessed or impoverished peasants and petty commodity producers into wage labour, or labour in search of wages. At first emigrants moved as individuals, families or groups and swelled the population of the gecekondu, but with the rapid expansion of opportunities for employment in Western Europe, parts of Arabia, and even Australia, these streams of migrants became rapidly transformed into an officially organised export of labour. By the mid-1970s, 800,000 workers had been sent abroad, and there was a substantial additional proportion of unrecorded and illegal migrants although estimates of this proportion vary. The official "queue" of those waiting for opportunities abroad was closed when it reached a further million names, and the mass phenomenon was abruptly halted by the reversal of immigration policies in Europe, following the the oil crisis of 1973. Officially sanctioned emigration, except for dependants, virtually ceased after 1975, although enterprising migrants have found opportunities elsewhere, such as Libya and Saudi Arabia.

The transformation of agrarian structures prior to 1970, then, was not just a process of mechanisation and commercial development. It was a complex process of differentiation and reorganisation of production, which involved a massive and extremely rapid uprooting of labour from the land, and the very uneven transformation of social relations of production.

Mass migration has had a number of effects on agriculture. Workers abroad who still have land, and family labour to work or supervise it, have been able to send remittances home to invest in land, machinery and increased production, but others have reduced household production to a bare subsistence or less, so that the survival of women, children and the elderly has depended on remittances, or removal to urban areas or Europe. Studies show that in general the productive effects of investment were minimal, except for the spread of tractors which have played a large part in creating a surplus labour force. It is difficult for small farmers to invest productively in household production, particularly from a distance. Penninx and Van Velzen describe Bogazliyan (one area of mass emigration) as "overmechanised", with human labour rendered redundant, although one wonders whether this includes women's work (Penninx and Van Velzen, in Abadan-Unat, et.al., 1976:303). State policies for organising the productive use of migrants' savings were singularly unsuccessful, particularly in the rural areas, so that migrants with savings to invest had to create additional outlets for themselves in urban areas (Ramazanoglu, 1977; Yasa, 1979; TUSIAD, 1983:96). Even where savings were brought to rural areas, there is some evidence that they were not invested locally (Penninx, Van Renselaar and Van Velzen, 1976).

Perhaps the most dramatic effect of mass emigration has been not so much changes introduced into peasant agriculture, but the situation that mass employment abroad for a limited period first removed surplus labour from the land, allowing capitalist agriculture to develop without hindrance, and then returned much of this labour to an agricultural system that could not absorb it.

The majority of early migrants were skilled or employed men from the most developed areas of Turkey. i.e. from existing wage labourers, and skilled workers, whom Turkey could least afford to lose (Paine, 1974; Miller and Cetin, 1973; Paine, 1974; Abadan-Unat et.al., 1976), but as mass migration developed, it drew in women, the unskilled, peasants from the estates of Eastern Turkey, and their various dependants, creating a vast pool of labour which was surplus to the needs of capitalist agriculture, which was not supported by a family wage, and which could barely be maintained by household production. This massive movement developed contradictions in the supply of cheap labour, and even led to local labour shortages in Turkey, (with some illegal import of labour from Pakistan and Bangladesh) and the loss of production in peasant agriculture which had to be compensated for by women and older people. The implications of these developments will be taken up below, but first I look more specifically at the pressures on peasant women brought about by labour migration.

2. The sexual division of labour in agriculture

The phenomenon of mass emigration from the rural areas in Turkey has been widely recognised as involving women as well as men, and has led to generalisations about the impact of migration on women, and examination of the extent of their "emancipation" (e.g. Abadan-Unat, 1977). There are studies which use survey data to look in very general terms at the status of Turkish women (e.g. Frey, 1963; Kudat Sertel, 1972; Fox, 1973a and 1973b; Kandiyoti, 1974; Ozbay, 1978) but the upheavals of Ottoman history, and the stormy birth of the Republic have led to most studies emphasising the great variety of Turkish social and economic life, which raises a number of questions about the exact nature of women's involvement not only in migration, but in the transformation of agrarian relations of production, of which migration is one aspect (Kiray, 1976; Kandiyoti, 1977).

In the transition from a Moslem empire to a secular republic, a rich variety of cultural traditions have been caught up with _Kemalist_ reforms, the development of Turkish capitalism, and the imposition of the Swiss Civil Code. This variety has meant considerable differences in the extent and manner of women's subordination to men, and in the sexual division of labour in agriculture. Although Turkish villages are often thought of as traditional strongholds of Islam, the content of each "tradition", the differences between Islamic sects, and the exercise of Islamic authority needs to be historically documented.

Turkish women were, as Abadan-Unat puts it, "catapulted into public life", wage labour and the nationalist struggle for independence, in the course of the crises precipitated by the final collapse of the Ottoman Empire and the outbreak of the First World War (Abadan-Unat, 1981:8). A few women even joined Ataturk's forces during the War of Independence (see e.g. Adivar, in Fernea and Bezirgan, 1978). Others are still almost totally subordinated to the service of their husbands and fathers-in-law. The changes of the twentieth century did not transform a shared pre-capitalist culture or a single system of production. Ottoman Turkey was characterised by great feudal estates, by nomadic herdsmen (_yoruk_), by peasant subsistence agriculture, by skilled crafts and so on. While women were generally subordinated, most women had productive roles, and even the most subordinated of girls could escape arranged marriages by institutionalised elopement (_kiz kacirma_) with the young man of their choice (Kudat, 1974; Starr, 1984). The impact of the uneven development of capitalism on such a differentiated society has brought changes in the sexual division of labour, and introduced class differentiation, so that new gulfs exist between, for example, the impoverished peasants of Eastern Turkey, the more privileged urban workers in the cities, and the rising bourgeoisie. Along with these sources of differentiation, Kemalist reforms brought new legal rights, secular state education and new social status for women, while capitalism has generated new forms of cultural and secular uniformity.

Given the complexity of these changes, it would seem most improbable that the impact of the development of capitalism on women's work could be a standard or general process, or that labour migration could always emerge in the same way. There are obvious differences between, for example: (a)the professional women of the bourgeoisie, who expect to combine full-time employment with child rearing and the management of a home, supported by female domestic servants, and private day nurseries, (b) women workers who enter factory labour forces and rely on female kin for help with domestic labour, (c) the "housewives" of the gecekondu whose apparent dependence on their husbands conceals an array of productive activities, and an army of domestic servants (Karpat, 1976; Kazgan, 1981; Senyapili, 1981), (d) peasant women left on the land by emigrant husbands, who take on the agricultural tasks of men as well as their own. Empirical studies of women's work in rural areas reveal some of the variety of the changes in women's lives, and in the sexual division of labour, brought about by the penetration of capitalism. Although such material is rather sparse in Turkey, they are studies which illustrate the dramatic impact of different aspects of the development of capitalism.

Bogazliyan: This district in the Yozgat province of central Anatolia was the site of a major investigation on the impact of mass migration, by a joint Dutch and Turkish research team. As part of this study, Yenisey describes the effects of migration on those who were left behind in two villages of the district. The population was split roughly equally between extended and nuclear households which acted as farming units. Emigrants from this area had been primarily men, although some had taken wives or daughters with them. Yenisey found two main changes in the local economy; an increase in consumption as a result of remittances from migrants, which raised the standard of living, but did not change the production system, and an increase in the labour input of those who remained. Where a wife and children were left on their own, women had taken over decision-making, relying on older children for help with domestic and agricultural tasks. In extended households, older household members were called on to keep the household economy going, so that migrants' mothers were often "unable to enjoy or to anticipate a restful old age", but it was the daughters-in-law of the household who took over the bulk of the work of the absent men. In one extended household the mother-in-law cared for the children, while the two daughters-in-law worked both in the fields and in the house. By eight at night the researchers found the two wives "asleep on their feet" (Yenisey, 1976: 335).

This example of absent men, and overworked women is one aspect of partial proletarianisation. Male workers, were not fully working class, in that they were not properly separated from their means of production, and used part of their wages to invest in household production. Women who remained in the villages were not fully peasants or farmers, since although they grew most of their own subsistence, and might sell some crops commercially, they used family labour and supplemented their income from migrants' wages.

Arpa: Brouwer and Priester studied the sexual division of labour in one village in the Konya province, in the mountains of central Anatolia, from which migrants had moved to Amsterdam. In the village there was a strict sexual division of labour, so that male emigration left tasks which were felt to be impossible or inappropriate for women. Women's work was seen as "domestic labour" but this included not only childcare, cooking, cleaning, carrying and heating water, and laundry, but also dressmaking and carpet weaving, which could bring in additional income, and a range of agricultural tasks, including raising food crops (which could be located as much as an hour's journey away from home) caring for livestock, making dairy products, helping to harvest cash crops, and cleaning and processing crops after harvest. The absence of men, not surprisingly, caused a drop in production.

Although some older women took over men's tasks, women generally argued that ploughing was men's work and women could not use ploughs or agricultural machinery. Women's normal work enabled women to feed their families ,and the sexual division of labour was sufficiently flexible for women to take over the grape harvest, juice extraction and drying of raisins, which was male work but non-mechanised. Even where these women lived in nuclear households they remained subject to the authority of their in-laws, and commercial transactions and the receipt of remittances was handled by fathers-in-law, or other male relatives of the husband. Given this situation, it is hard to see what women could have gained by taking over more male tasks. Brouwer and Priester note that rather than fighting for the right to do men's work, in addition to their own, what women wanted was to join their husbands in Amsterdam, and there was some evidence that women were struggling to gain control of their husband's remittances. Although some women had joined their husbands, family reunion was difficult and expensive since women could not obtain work permits for some time. Men's interests were better served if they could find European women with whom to set up house abroad, while retaining a wife who could provide subsistence for herself and their children in the village, under the supervision of himself on his annual holiday, and his male kin in the village. The authors argue strongly against the view that either taking over men's work in the village or wage labour in Europe brings any real change in the subordination of women (Brouwer and Priester, 1983).

In this study it becomes clear that migration is not only an aspect of partial proletarianisation, it is also a process that can continue or intensify the subordination of women, both in the village and in the cities. A husband's absence brought little real freedom if extended kinship groups retained patriarchal control. Similarly, work in Europe brought about the partial proletarianisation of women, but when they migrated as wives or daughters, they still worked as subordinated members of a family group.

Alihan and Yenikoy: Engelbrektsson studied two village, also in Konya province, from which migrants had gone to Sweden. These villages, however, differed from each other considerably in terms of "traditional" culture and the relations between the sexes. Alihan which was a village of poor landowning peasants had approximately 135 emigrants (who had gone to six countries between them) from a population of around 650, of whom only 15 were women (all of them wives of male emigrants). Yenikoy, was a comparatively wealthy village with most households receiving good incomes from mechanised farming. Yenikoy had only 28 emigrants from a population of around 900, of whom most were married couple from the poorest kin group.

When these women reached Europe, the women from Alihan (with the exception of one whose husband had a drinking problem) did not work outside the home, and so became "expensive" dependants. The women from Yenikoy, on the other hand, were taken to Europe in order to be put to work, but had exercised considerable influence on the decision-making process, and were regarded as sufficiently responsible to be able to manage outside the home without constant male supervision. Nevertheless their lives abroad were very circumscribed, as they remained within their kinship groups, and their earnings were handed over to their husbands. A major change had come in the efforts of younger men to separate themselves from the control of their fathers. If this change is achieved in future, women perhaps have more chance of struggling for control of their own resources against their husbands, than they do against a phalanx of husband and male in-laws. Greater autonomy for younger nuclear families will also have consequences for the continued attachment of wage earners to the land, since absentee agricultural production is greatly dependent on the co-operation of extended kin groups (Engelbrektsson, 1978).

Brouwer and Priester point to several similarities between their findings and those of Engelbrektsson. They also suggest that their findings on the sexual division of labour could be generalised to "rural Turkish society as a whole". From the questions raised by Engelbrektsson, and from the other studies discussed here, however, it seems very unlikely that this could be the case. The subordinated daughters-in-law who take over their husbands' work and and submit to the economic and social control of their parents-in-law without any struggle to preserve their own autonomy, represent only one type of response to male emigration, and only one type of migration.

Bodrum: Starr in a rather different study in a much more developed region of Aegean Turkey, argues that capitalism does not necessarily affect the position of women adversely. Starr suggests that the development of capitalist agriculture in the Bodrum region produced a class system which made intra-village marriages an advantageous means of consolidating landholdings. This form of marriage served to protect women from the control of their in-laws, by keeping them close to their own female kin, while the access to the law granted to women in the Republic has enabled women to use the law effectively to exercise their land rights. Access to land in their own right gave at least some women the opportunity to become small capitalist farmers. The disruptive effects of the emergence of capitalist agriculture and the development of private property were counteracted to some extent by equality before the law. The introduction of tangerines as a cash crop encouraged the gaining of land titles and investment in agriculture, in which women played an active part. Starr gives here one detailed case study of the process of differentiation which capitalist transformation can bring (Starr, 1984).

The advent of mass migration in Turkey has transformed the world of peasant women. While women have always worked, the penetration of capitalism into agriculture has meant more work for women, but also considerable differentiation between women. Where women remained on the land as wives of absent workers, they were under pressure to work harder, to move into wage labour or other work in the cities or abroad or, if they could not maintain themselves, to become dependants for the first time. Where they entered wage labour they tended to find "women's work" or sweated labour at the bottom of the labour market, where they lacked any basis of class solidarity with male workers. At the same time, the independence of wage earning was mediated by their family membership and their continued subordination within their husband's kinship group.

There is no logic in the development of capitalism that demands that workers be male. Wherever men are unavailable (for example through war) or are seen as too expensive, politically unreliable or otherwise stereotypically unsuitable, women have been found to have the necessary qualities for whatever work is needed. The choice of who migrates is then largely a matter of power and ideology tempered with practical considerations, such as the prevailing division of labour. [5] In Turkey the choices made at village level seem to be generally controlled by groups of male kin, with women sometimes totally subordinated, sometimes being involved in decision-making, and sometimes becoming relatively independent. This has meant that migration was initially mainly a male phenomenon, but that where women could exercise choice, or where women were seen as too weak to be left without their husbands or fathers, they could migrate in their own right or as dependants (many of whom could eventually become workers). Women migrants have included, therefore, both the most independent of Turkey's working women, and the most subordinated of peasant wives. Such broad ideological factors as Islam, and patriarchal attitudes, mould the expectations of potential workers and employers, but these prove suitably adaptable when labour shortages develop.

Once in urban areas, women can become isolated within their nuclear families and solely responsible for the care of their children. In other parts of the world where it is normal for single women to migrate to urban employment, women have more independent access to wage employment, more control of reproduction and sole responsibility for their children. The situation for Turkish women is varied, but women still generally marry, and marriage entails subordination. There are quite dramatic differences in class interests between peasant and working women, and bourgeois women. Bourgeois women have made little political impact, but unlike the first bourgeois women in Europe, they are active in public life, in the professions, banking and service industry, and exercise their legal rights to property ownership. Behind the general "causes" of migration, lie rather different consequences for women's work, for the transformation of the social organisation of production and for the incorporation of men and women into the working class. Migration cannot simply be seen as men's response to poverty, underdevelopment or excess population. Labour migration in Turkey has entailed the continuing subordination of women, the flexibility of Islamic culture and customs, and rather different struggles of women in the production and reproduction of their lives. [6]

3. The emergence of classes: incomplete proletarianisation and state intervention

While it is clear that the development of Turkish capitalism required a fundamental transformation of agrarian structures; in particular, land reform, the reorganisation of pricing, irrigation, mechanisation, and the development of a wage labour force, these complex requirements generated serious contradictions, and government policies have rarely achieved their intended effects (Hershlag, 1968; Ozbudun and Ulusan, 1980). These points become crucial in accounting for Turkish labour migration at this period, since the relation between the activities of the state, the interests of industrial and agricultural capitalists, the interests of capitalists elsewhere, and the exact supply of labour are neither constant nor harmonious.

The supply of unskilled labour withdrawn from the land did perhaps work in the interests of the large landowners and wealthier farmers who could afford to mechanise their farms; it did work in the interests of those peasants who could enter the formal working class or petty bourgeoisie and improve their skills and standard of living. [7] The capitalist class nationally and internationally did presumably have their interests directly met by the impoverishment of potential workers who could act as a reserve army of labour, but contradictions were inherent in this labour supply. Surplus labour was soon seen by the states of Europe, as well as of Turkey, as an immediate political threat rather than as a flexible economic resource. While this process of creating surplus labour has occurred in recent years in all parts of the world, the manner of its creation, and the consequences for the state have varied, and we need accounts of these variations in order to understand the process in general.

The accelerated development of wage labour in Turkey coincided with two other developments; the end of the industrial boom in Europe, following the oil crisis of 1973, and the end of Turkey's boom period of import-substitution. The combination of these historical events shaped the manner of state intervention in the creation of migrant labour, and contributed directly to the intensification of class struggles in Turkey, in which labour migration has been one contributory factor. These two boom periods came to an end during the 1970s, precipitating Turkey directly into the period of economic and social crisis which led to the coup of 1980. This period is analysed in other chapters in this book, but in order to understand the place of labour migration in relation both to the experiences of individual migrants, and in relation to the uneven development of Turkish capitalism, we need to examine the impact of the coincidence of the end of import-substitution and the end of the mass export of labour.

During the 1960s and early 1970s. Turkey had been able to export labour because of growing demand in the more developed parts of Europe. Where labour was imported, the receiving countries initially gained labour that was often skilled, and also cheap in that it was produced, serviced and reproduced in Turkey (see Wolpe, 1972; Castells, 1975; Burawoy, 1976). Some sectors of European industry, however, soon became structurally dependent on imported "temporary" labour and on skilled labour. This dependence created considerable problems of how to retain essential workers and stabilise the skilled labour force, while retaining a flexible and docile supply of unskilled labour. The needs of employers for both stable and fluctuating labour forces created political problems for the state, and competition with domestic labour which generated well-documented contradictions which have been simmering in Europe ever since (Castles and Kosack, 1973; Castles, 1984; Marshall, 1973; Ward, 1975a and 1975b; Krane, 1975; Phizacklea, 1983).

In common with migrants from other areas, Turks who crossed national boundaries were subjected to two states, with the consequent loss of political rights and class alliances in both, which affected class solidarity and class practices at home and abroad. Contradictions in the process of proletarianisation, therefore, can be conceived as developing both within the Turkish social formation and within the labour-importing social formations.

After 1973, the onset of recession in Europe not only halted further recruitment, but also heightened awareness of the presence of a large and increasingly resident foreign labour force. Once new immigration ceased, those who had work in Europe were unable to move freely back and forth between home and work, and so had a new incentive to try to settle,and good reason to bring their spouses, parents and children out of Turkey. This development had a number of consequences. While immigrant workers (notably in West Germany, the major employer of Turkish workers) were politically subordinated and served to fragment working class consciousness, they became increasingly integrated into a labour-intensive industrial system, and also more "expensive". At its peak, mass emigration drew in women and older workers, and as men settled abroad, wives, children and parents joined them. The reproduction and maintenance of the labour force thus began to shift from Turkey to Europe with demands being made for health care, family housing, education and leisure facilities. These demands remained relatively limited but took on a new political significance as the recession deepened and immigrants were labelled as an expensive problem group competing with natives for scarce resources.

Immigrant labour is only "cheap" when free movement between wage labour and household production is permitted. Once workers are forced to settle and thus encouraged to bring in "dependants", permanent proletarianisation is encouraged with the need for a family wage. The political interests of the state can come into direct conflict with the economic interests of employers, but the issues are unlikely to be clear cut. The distinct ethnic identity of immigrant workers serves to prevent the development of working class consciousness, and low wages encourages the most exploitative forms of women's work (see for example, Phizacklea, 1983).

During the boom period of state-sponsored emigration, the impact of migration within Turkey and abroad differed. While both groups of migrants sent remittances home to maintain their dependants, or to invest in agriculture or small businesses, remittances sent from abroad had the effect of bringing in much needed foreign exchange, and of increasing spending power within Turkey. Since industry was based on import-substitution, this increased spending power served to expand the domestic market for Turkish goods and thus encouraged the profitability of import-substitution. Although Turkish workers accumulated savings in European banks, remittances flowed back to Turkey on a massive scale. During 1971-72, foreign exchange flows from remittances exceeded those from exports by 50 per cent (Pamuk, 1981). Between 1969 and 1982 a sum equivalent to $14.3 billion was injected into the Turkish economy by Turks working abroad (<u>TUSIAD</u>, 1982:97) and for a short period before the recession deepened, Turkish workers financed a widening and ultimately disastrous trade gap.

The extent to which Turkish workers have managed to settle abroad and yet to maintain their links with household production or small entrepreneurial activities, is indicated by the scale of these remittances. Proletarianisation remains partial where these links are still maintained, and this area need further investigation.

In Turkey, the availability of employment abroad and this inflow of foreign exchange from migrants' remittances, meant that the state, far from attempting to control the withdrawal of labour from the land, actively encouraged the transformation of the peasantry into wage labour in a movement which had got out of hand even before the European nations closed their borders to new immigrants. The short term gain of foreign exchange which masked Turkey's acute balance of payments shortfall, meant that when the flow of remittances faltered, Turkey had no substitute source of foreign exchange. By 1978, the lack of foreign exchange was critical, and Turkey's mainly import-substitution industry was operating at around a third of its capacity. The chickens of opportunist emigration policies came home to roost with the IMF.

Between 1962 and 1977, Turkey's import-substitution policies led to a boom in domestic consumer industries producing for the home market, and heavily protected by the state. The import of intermediate and capital goods which were needed to set up these industries, were financed largely by agricultural exports and workers' remittances (TUSIAD, 1978). The contradictions generated by these policies at this period developed into the crisis discussed in other chapters in this book. Easy profits from highly protected domestic industry discouraged investment by the industrial bourgeoisie in the capital goods industry and made competitive exports unnecessary.

This policy of import-substitution has been tried in other parts of the world, notably Latin America. but it is one that by its nature cannot continue indefinitely. Once the domestic market is saturated with the available consumer goods, and cannot be further expanded under prevailing conditions, some radical change is required. Since the nature of this change requires state intervention in resolving competing interests, the outcomes of the changes that follow are not necessarily the same. As Pamuk argues:

> "The outcome will depend on the balance of internal class forces, the nature of ties to the international capitalist system and the world conjuncture."
> (Pamuk, 1981, p.28)

In spite of making this point, and also pointing out that the outcome of the new "economic miracle" in Brazil has been different from the development of events in Turkey, Pamuk is constrained from developing his point further since he conceives the Turkish case as typical of "peripheral capitalism".

The insistence on Turkey as typical of peripheral capitalism, serves to limit the questions which can be asked about what was happening during the 1970s, in the capitalist world system as a whole, and about the differences between Turkey and other cases of import-substitution. It can be said in general that import-substitution offers a short term solution to certain common problems among late-industrialising social formations, but that it cannot be depended on as a long term strategy within an unequal international division of labour.

Although capitalist production always creates the need for wage labour, this does not mean that it was always clearly in the interests of the Turkish capitalist class to produce so suddenly a massive surplus labour force which could not all be absorbed by non-agricultural employment. While it is true that the economies of Western Europe benefitted for a while from "cheap" labour that was reproduced elsewhere, it is questionable as to how far it was in the long term interests of European states to import a massive "temporary" labour force which became increasingly permanent, and how far it was in the interests of the Turkish state to pull yet more labour from the land in order to earn foreign exchange and depress wages. The existence of welfare provision in Europe makes any unproductive labour relatively costly to maintain, while the lack of welfare provision for surplus labour in Turkey creates a volatile situation of destitution and subsistence, which can directly threaten the stability of the state. In Europe, as in Turkey, meeting the short term interests of industrial capital, can run counter to the interests of other dominant groups represented in the state; notably agricultural and commercial capital. Mass emigration brought Turkey much needed foreign exchange, and the economies of Europe a brief period of cheap labour, but the long term social, economic and political problems which have been created are immense.

The Turkish state was directly and indirectly involved in drawing labour off the land, but state intervention is not necessarily successful just because attempts to intervene are made. The state has, for example, made efforts to drive workers back to the land with conspicuous lack of success. In fact it seems generally true that labour will not return to household production unless either household production is further transformed by capitalist penetration (see for example, Hansen and Marcussen, 1982) or considerable force is used (of which South Africa is an obvious example). State intervention in the transformation of agricultural relations of production was extremely effective in Turkey, because of the combination of circumstances which occurred at this particular period, but effective intervention did not mean that subsequent developments unfolded in the desired manner. Birtek and Keyder have argued that the agricultural sector presented the state with major obstacles to its attempts to guide the economy. The institutional framework governing the allocation of resources between sectors, was dependent on shifting alliances within and between the emerging classes. The state could not simply act for the bourgeoisie since the bourgeoisie represented different interests in agricultural change (Birtek and Keyder, 1975). The state itself embodied contradictory interests with regard to the use of labour, and further contradictions were generated as labour left the land in increasing numbers.

The consequences of the end of import-substitution in Turkey have been the products of particular historical circumstances, and a particular stage in the development of classes and the state in Turkey. When the end of the boom period of import-substitution coincided with the ending of mass emigration and the reduction of migrants' remittances, [8] this combination of circumstances brought about a critical shortage of foreign exchange, rocketing inflation and unemployment, accompanied by appeals to the West for rescue, and short term borrowing in international markets on unfavourable terms (Pamuk, 1981:29). These developments were accompanied by the onset of recession in Europe, leading to the return of workers from abroad. Some of these returned because of unemployment, but others chose to return in the hope of setting up business or commercial farms in Turkey before employment abroad became impossible. These were generally unable to invest their savings productively or to find employment and were caught up in a subsequent period of extreme and complex forms of class struggle fought out in the education system, the shanty towns, and the rural areas with considerable loss of life and eventually, the destruction of the Turkish left.

While state intervention in creating and attempting (usually unsuccessfully) to control labour migration is common, the outcomes of such attempts are variable. (Compare, for example, the effectiveness of state control in South Africa, with that of India). Contrary to Wallerstein's astonishing view that a struggle for control of the state structure does not occur within the peripheral economies of the world system (Wallerstein, 1979:200), the Turkish case shows clearly that the men, women and children who were encouraged or driven from their villages and herded abroad in their thousands were caught up in a complex set of processes emanating from a desperate and violent domestic struggle for control of the state, and through state power for control of the direction which the uneven development of Turkish capitalism could take.

CONCLUSION

Stressing the specificity of the social formation, and thus of the causes and consequences of labour migration, is often seen as of little interest. Everyone knows individual cases are unique, but the social theorist seeks to grasp these cases within general theory. In this chapter I have argued that in order to understand the nature of labour migration in general and what happened in Turkey in the 1970s, it is not relevant to ask whether or not Turkish migration is more or less typical of peripheral capitalism than say Brazilian or Ugandan movements of labour. It is more illuminating to ask whether conceptualising work-seeking peasants in terms of the the transformation of social relations of production in agriculture, the changing sexual division of labour and contradictory processes of class formation and struggle, helps explain the place of labour migration in the development of capitalism, both generally and in specific instances. Although a number of detailed studies of emigration, return migration and their consequences have been made in Turkey, empirical analysis of the emergence of capitalism and the transformation of agrarian structures is relatively little developed. This chapter, therefore, raises more questions than it is able to answer, and much work remains to be done.

Turkey's policy of encouraging the emigration of labour, however, cannot be dismissed as the mass breakdown of rural socialisation, or as the simple transfer of surplus population from one region to another. It has to be seen as a short term and desperate measure at a particular stage in the development of Turkish capitalism, at a particular stage in the incorporation of the Turkish economy into the world capitalist system, and at a particular stage in the complex processes of class formation, and the shifting of class alliances in Turkey. The scale of migration reached by the 1970s indicates the state's loss of control over the transformation of agrarian structures and the development of industry, the failure of any one fraction or alliance of the Turkish bourgeoisie to dominate state power, and the failure of the industrial bourgeoisie to develop heavy industry. The resulting intensification of uneven development, and the pauperisation of a large proportion of the population, was directly threatening the interests of the state. It was this threat to the state, among other factors, which led to military intervention in 1980, in order to preserve social order and to enforce the shift to an outward-looking economy.

Seeking to conceptualise migration in these terms points to the links between Yenisey's exhausted daughters-in-law, through a series of interrelated events and processes to the decision of the military to seize power in 1980 and implement the IMF stabilisation programme, within the general framework of constraints imposed by Turkey's position within the world capitalist system.

Although capitalism is clearly the dominant mode of production, in the Turkish social formation, this dominance has not resulted in the emergence of classes in anything approaching a pure form. [9] Labour migration is one complex aspect of a complex and contradictory process of class formation. Where men work in German factories, and their wives and children labour on their own land for their own subsistence, or hire themselves to a landowner for seasonal harvesting, proletarianisation is in evidence, but is only partially achieved. Where landowners are developing commercial crops but depending on "feudal" labour, the emergence of an agricultural capitalist class is in process rather than completed. Where differentiation among the peasantry enables some to use wages to invest in mechanised profit-making farming, drawing on family labour, or using dispossessed kin or neighbours as labour or share-croppers, processes of class formation are in motion but there is no clear or completed transition to capitalism. Even where peasants have moved to Europe and appear unequivocally proletarian, the volume of remittances still sent to Turkey indicates continuing ties both with the land and with small entrepreneurial activities. When factory or construction workers who may have been class conscious union members in Europe return to Turkey, they may shift back into household production, or into petty commodity production or shopkeeping. In terms of class position men and women in the same family can stand in different relations to the production system, thus complicating the situation even further. Some authors, (e.g. Rudra, 1978; Kitching, 1980), have argued in effect that if classes cannot be identified at a given moment in time, then class analysis must be irrelevant to understanding the development of capitalism. This static conceptualisation of class, however, successfully mystifies the essentially dynamic and contradictory nature of labour migration.

Labour migration is nothing new in the history of capitalism, but it is constantly developing new contradictions, new complexities and new consequences. If we start by looking for changes in the ways in which agricultural production is organised in Turkey, labour migration can be seen as part of a great and messy upheaval, as a very uneven process of transformation gets under way. It also becomes clear that while Turkey remains disadvantaged as a late-industrialising social formation in a highly competitive world system, the prospects for a fuller transformation must be seriously constrained.

NOTES

1. Parts of this paper were originally incorporated into "The Articulation of Modes of Production, Class Struggles and Labour Migration" written in collaboration with Janet Bujra and Huseyin Ramazanoglu, and presented to the British Sociological Association's Annual Conference, Aberystwyth, 1981. I am grateful to Janet Bujra and to Huseyin Ramazanoglu for comments on this chapter, and to Raymond Apthorpe for comments on my intentions.

2. In this chapter I have drawn on fieldwork carried out in Istanbul in 1975, funded by a grant from the then Social Science Research Council (Ramazanoglu, 1977), and on subsequent reconsideration of the theoretical framework originally adopted.

3. These issues are being raised directly in work currently being carried out by Diane Elson and Ruth Pearson, and arise in part from their earlier work (Elson and Pearson, 1981).

4. I am indebted to Ronnie Frankenberg for admonishing me with this text.

5. As Sassen-Koob has argued, powerlessness is an important characteristic of desirable labour (Sassen-Koob, 1980).

6. There are very considerable differences within the Moslem Middle East in the ability of peasant women to act independently, and to engage in commercial and legal transactions. There are also class differences, but this does not mean that bourgeois women do not experience subordination. These differences need careful exploration, which cannot be attempted here, although the issue is directly relevant to a full understanding of labour migration.

7. There are a number of detailed studies on the effect of migration on wage levels and per capita income, which cannot be reviewed here. There is some evidence that while individual peasants can improve their incomes through migration, and village incomes can be raised through receipt of migrants' remittances, the long term effects of mass migration has a depressing effect on wages (Kolan, 1976; Kuran, 1980).

8. In absolute terms, migrants' remittances increased from 1979. Although new workers were not being actively sought, workers and dependants were still going abroad, and workers settled abroad still supported their kin in Turkey. There is some evidence that remittances per head dropped, and allowance must be made for inflation. These trends are documented in the annual TUSIAD reports on The Turkish Economy.

9. The very complex problems of establishing empirically the domination of capitalism in a given social formation are discussed in the debate on the Indian Mode of Production (see for example, Rudra, et. al., 1978).

REFERENCES

Abadan-Unat, Nermin (1974); "Turkish External Migration and Social Stability", in P. Benedict, E. Tumertekin and F. Mansur (eds.), Turkey: Geographical and Social Perspectives, Brill, Leiden

Abadan-Unat, Nermin; Keles, Rusen; Penninx, Rinus; van Renselaar, Herman; van Velzen, Leo; Yenisey, Leyla (1976); Migration and Development: A Study of the Effects of International Labour Migration on Bogazliyan District, Ajans-Turk Press, Ankara

Abadan-Unat, Nermin (1976); Turkish Workers in Europe 1960-1975: A Socio-Economic Reappraisal, Brill, Leiden

Abadan-Unat, Nermin (ed.) with D. Kandiyoti and M. Kiray (1981); Women in Turkish Society, Brill, Leiden

Adivar, Halide Edip (1978); "Excerpts from Memoirs and the Turkish Ordeal", in E. Warnock Fernea and B. Qattan Bezirgan, Middle Eastern Moslem Women Speak, University of Texas Press, Austin

Akello, Grace (1982); Self Twice Removed: Ugandan Women, Change-International Reports: Women and Society, London

Aker, Ahmet (1972); Isci Gocu (Labour Migration), Sander Yayinevi, Istanbul

Berberoglu, Berch (1982); Turkey in Crisis, Zed Press, London

Birtek, Faruk and Keyder, Caglar (1975); "Agriculture and the State: An Inquiry into Agricultural Differentiation and Political Alliances: The Case of Turkey", Journal of Peasant Studies, Vol.2, No.4, pp.446-67.

Brouwer, Lenie and Priester, Marijke (1983); "Living in Between: Turkish Women in their Homeland and in the Netherlands", in A. Phizacklea (ed.), One Way Ticket: Migration and Female Labour, Routledge & Kegan Paul, London

Buroway, Michael (1976); "The Function and Reproduction of Migrant Labour; comparative Material from Southern Africa and the United States", American Journal of Sociology, Vol.LXXXI, No.5, pp.1050-87

Bryceson, Deborah F. (1980); "The Proletarianisation of Women in Tanzania", Review of African Political Economy, No.17, pp.4-27.

Bujra, Janet (1977); "Production, Property, Prostitution: Sexual Politics in Atu", Cahier d'Etudes Africaines, Vol.XVII, pp.13-40.

Bujra, Janet (1983); Urging Women to Redouble their Efforts: Class, Gender and Capitalist Transformation in Africa, BSA Conference Paper, mimeo.

Castells, Manuel (1975); "Immigrant Workers and Class Struggles in Advanced Capitalism", Politics & Society, Vol.5, No.1, pp.33-66

Castles, Stephen and Kosack, Godula (1973); Immigrant Workers and Class Structure in Western Europe, Oxford University Press/Institute of Race Relations, Oxford

Castles, Stephen (1984); Here to Stay, Pluto, London

Cliffe, Lionel (1979); "Labour Migration and Peasant Differentiation: Zambian Experiences", in B. Turok (ed.), Development in Zambia: A Reader, Zed Press, London

Deere, Carmen D. and de Leal, Magdalena L. (1982); "Peasant Production, Proletarianisation and the Sexual Division of Labour in the Andes", in L. Beneria (ed.), Women and Development, Praeger, New York

Eberhard, Wolfram (1954); "Change in Leading Families in Southern Turkey", Anthropos, Vol.XLIX, pp.992-1003

Elson, Diane and Pearson, Ruth (1981); "Nimble Fingers Make Cheap Workers: An Analysis of Women's Employment in Third World Export Manufacturing", Feminist Review, No.7, pp.87-107

Engelbrektsson, Ulla-Britt (1978); The Force of Tradition: Turkish Migrants at Home and Abroad, Acta Universitatis Gothoburgensis, Gothenburg

Fay, Brian (1974); Social Theory and Political Practice, Allen & Unwin, London

Fox, G.L. (1973a); "Some Determinants of Modernism among Women in Ankara, Turkey", Journal of Marriage and the Family, Vol.XXXV, No.3, pp.520-29

Fox, G.L. (1973b); "Another Look at the Comparative Resources Model: Assessing the Balance of Power in Turkish Marriages", Journal of Marriage and the Family, Vol.XXXV, No.4, pp.718-30

Frey, Frederick (1963); "Surveying Peasant Attitudes in Turkey", Public Opinion Quarterly, Vol.XXVII, No.3, pp.335-55

Foucault, Michele (1982); "Afterword", in H. Dreyfus and P. Rabinow, Michel Foucault: Beyond Structuralism and Hermeneutics, Harvester, Brighton

Hansen, M.B. and Marcussen, H.S. (1982); "Contract Farming and the Peasantry: Case Studies from Kenya", Review of African Political Economy, No.23, pp.9-36

Hershlag, Z.Y. (1968); Turkey: The Challenge of Growth, Brill, Leiden

Hinderink, Jan and Kiray, Mubeccel (1970); Social Stratification as an Obstacle to Development: A Study of Four Turkish Villages, Praeger, New York

Hirsch, Eva (1970); Poverty and Plenty on the Turkish Farm: A Study of Income Distribution in Turkish Agriculture, Columbia University Press, New York

Johnson, Carlos (1979); "Critical Comments on Marginality: Relative Surplus Population and Capital/Labour Relations", Labour, Capital & Society, Vol.12, No.2, pp.77-112

Kandiyoti, Deniz (1974); "Some Social-Psychological Dimensions of Social Change in a Turkish Village", British Journal of Sociology, Vol.XXV, No.1, pp.47-62

Kandiyoti, Deniz (1977); "Sex Roles and Social Change: A Comparative Appraisal of Turkey's Women", Wellesley Editorial Committee, Women & Development, Chicago University Press, Chicago

Karpat, Kemal (1960); "Social Effects of Farm Mechanisation in Turkish Villages", Social Research, Vol.XX, No.7, pp.83-103

Karpat, Kemal (1976); The Gecekondu, Cambridge University Press, Cambridge

Kazgan, Gulten (1981); "Labour Force Participation, Occupational Distribution, Education Attainment and the Socio-Economic Status of Women in the Turkish Economy", in N. Abadan-Unat (ed.) with D. Kandiyoti and M. Kiray, Women in Turkish Society, Brill, Leiden

Keles, Rusen (1976); "Investment by Turkish Migrants in Real Estate" in N. Abadan-Unat (ed.), Turkish Workers in Europe 1960-1975, Brill, Leiden

Kiray, Mubeccel (1976); "The Family of the Immigrant Worker in Turkey", in N. Abadan-Unat (ed.), Turkish Workers in Europe 1960-1975, Brill, Leiden

Kiray, Mubeccel (1981); "The Women of Small Town", in N. Abadan-Unat (ed.) with D. Kandiyoti and M. Kiray, Women in Turkish Society Brill, Leiden

Kitching, Gavin (1980); Class and Economic Change in Kenya, Yale University Press, New Haven

Kolan, Tufan (1976); "An Analysis of Individual Earning Effects Due to External Migration", in N. Abadan-Unat (ed.), Turkish Workers in Europe 1960-1975, Brill, Leiden

Krane, Ronald E. (ed.), (1975); Manpower Mobility Across Cultural Boundaries: Social Economic and Legal Aspects: The Case of Turkey and West Germany, Brill, Leiden

Kubat, Daniel (1979); "Turkey", in D. Kubat with U. Merklander and E. Gehmacher (eds.), The Politics Migration Policies, Center for Migration Studies, New York

Kubat, Daniel and Hoffman-Nowotny H-J. (1982); "International and Internal Migration: Towards a New Paradigm", in T. Bottomore, S. Nowak, M. Sokolowska, Sociology: The State of the Art, Sage, Beverly Hills

Kudat, Ayse (1974); "Institutional Rigidity and Individual Initiative in Marriages of Turkish Peasants", Anthropological Quarterly, Vol.XLVVI, No.3, pp.288-303

Kudat, Ayse (1975); "Sociological Impacts of Turkish Migration", Etudes Migrations, Vol.XII, No.38-39, pp.330-41

Kudat, Ayse (1982); "Personal, Familial and Societal Impacts of Turkish Women's Migration to Europe", in UNESCO, Living in two Cultures: The Socio-cultural Situation of Migrant Workers and their Families, Gower/UNESCO Press, London

Kuran, Timur (1980); "Internal Migration: The Unorganised Urban Sector and Income Distribution in Turkey 1963-1973", in E. Ozbudun and A. Ulusan (eds.), The Political Economy of Income Distribution in Turkey, Holmes & Meier, New York

Marshall, Adriana (1973); The Import of Labour: The Case of Netherlands, Rotterdam University Press, Rotterdam

Marx, Karl (1970); A Contribution to the Critique Political Economy, Progress Publishers, Moscow

Marx, Karl (1976); Capital, Vol.1, Penguin, Harmandsworth

Matras, Judah (1973); Populations and Societies, Prentice-Hall, New York

Meillassoux, Claude (1981); Maidens, Meal and Money: Capitalism and the Domestic Community, Oxford University Press, Oxford

Miller, Duncan and Cetin, Ihsan (1974); Migrant Workers and Labour Markets, Institute of Economic Development, Istanbul University, Istanbul

Munro, John H. (1974); "Migration in Turkey", Economic Development and Cultural Change, Vol.XVII, No.4, pp. 634-53

O'Laughlin, Bridget (1977); "Production and Reproduction: Meillassoux's Femmes, Greniers et Capitaux" Critique of Anthropology, No.8, pp.3-32

Omvedt, Gail (1980); "Migration in Colonial India: The Articulation of Feudalism and Capitalism by the Colonial State", Journal of Peasant Studies, Vol.7, No.2, pp.185-212

Ozbay, Ferhunde (1978); Changing Women's Position in the Family in Turkish Villages, Institute of Population Studies, Hacettepe University, Series ISA, Ankara

Paine, Suzanne (1981); Exporting Workers: The Turkish Case, Cambridge University Press, Cambridge

Pamuk, Sevket (1981); "The Political Economy of Industrialisation in Turkey", MERIP Reports, No.93, pp.26-32

Peek, Peter and Standing, Guy (eds.) (1982); State Policies and Migration: Studies in Latin America and the Caribbean, Croom Helm, London

Penninx, Rinus, van Renselaar, H., van Velzen, Leo (1976); Social and Economic Effects of External Migration in Turkey, R.E.M.P.L.O.D., I.D.S., The Hague (mimeo)

Phizacklea, Annie (ed.) (1983); One Way Ticket: Migration and Female Labour, Routledge and Kegan Paul, London

Ramazanoglu, Caroline (1977); Return Migration and Investment: Some Theoretical, Empirical and Planning Problems from a Study of Returned Migrants in Istanbul, S.S.R.C. Report (mimeo)

Rudra, Ashok (1978); "In Search of the Capitalist Farmer", in A. Rudra, et.al., Studies in the Development of Capitalism in India, Vanguard Books, Lahore

Rudra, Ashok, et.al. (1978) Studies in the Development of Capitalism in India, Vanguard Books, Lahore

Sassen-Koob, Saskia (1980); "Immigrant and Minority Workers in the Organisation of the Labour Process", The Journal of Ethnic Studies, Vol.VIII, No.1, pp.1-34

Senyapili, Tansi (1981); "A New Component in Metropolitan Areas. The

Gecekondu Women", in N. Abadan-Unat, et.al., (ed.), Women in Turkish Society, Brill, Leiden

Sertel, Ayse Kudat (1972); "Sex Differences in Status and Attitudes in Rural Turkey", Hacettepe Bulletin of Social Sciences and Humanities, Vol.IV, No.1, pp.48-79

Standing, Guy (1979); "Migration and Modes of Exploitation", WEP 2-21/wp 72, ILO (mimeo), Geneva

Standing, Guy (1980); "Semi-Feudalism, Migration and the State in Guyana", WEP 2-21/wp 73, ILO (mimeo), Geneva

Starr, June (1984); "The Legal and Social Transformation of Rural Women in Aegean Turkey" in R. Hirschon (ed.), Women and Property: Women as Property, Croom Helm/St Martin's Press, London

Stirling, Paul (1965); Turkish Village, Wiley, New York

TUSIAD (Turkish Industrialists and Businessmen's Association) (1978); The Turkish Economy, TUSIAD, Istanbul

Tutengil, Orhan (1972); "Small Village and Large City: Problems Regarding Socio-Economic and Cultural Development of Turkey", in Problems of Turkey's Economic Development, Vol.1, Institute of Economic Development, Istanbul University, Istanbul

Ulusan, Aydin (1980); "Public Policy Towards Agriculture and its Redistributive Implications", in E. Ozbudun and A. Ulusan (ed.), The Political Economy of Income Distribution in Turkey, Holmes & Meier, New York

Wallerstein, Immanuel (1979); The Capitalist World Economy, Cambridge University Press, Cambridge

Ward, Anthony (1975a); "European Capitalism's Reserve Army", Monthly Review, Vol.XXVII, No.6, pp.17-32

Ward, Anthony (1975b); "European Migratory Labour: a Myth of Development", Monthly Review, Vol.XXVII, No.7, pp.24-36

Wolpe, Harold (1972); "Capitalism and Cheap Labour Power in South Africa", Economy and Society, Vol.I, No.4, pp.425-56

Yasa, Ibrahim (1979); Yurda Donen Isciler ve Toplumsal Degisme (Returning Workers and Social Change), Turkiye ve Orta Dogu Amme Idaresi Enstitusu, Ankara

Yenisey, Leyla (1976); "The Social Effects of Migrant Labour on the District Left Behind: Observations in Two Villages of Bogazliyan" in N. Abadan-Unat et.al., Migration and Development: A Study of the Effects of International Labour Migration on Bogazliyan District, Ajans-Turk Press, Ankara

Zedong, Mao (1967); Selected Works, Vol.1, Foreign Languages Press, Peking

7 Military intervention and the crisis in Turkey*

FEROZ AHMAD

According to Military Communique No.1 broadcast at about 6 o'clock in the morning, local time, the causes of the military intervention in Turkey were as follows: the state and its principal organs had been rendered inoperative, the constitutional structure was full of contradictions, and political parties were intransigent in their attitude and lacked the consensus necessary to deal with the country's problems. As a result of all of these factors, secessionist forces had increased their activities and the life and property of the citizens were no longer secure. The reactionary and other deviationist ideologists, read the communique, flourished instead of Ataturkism or Kemalism. Attacks on all aspects of society - schools, the universities, the judiciary, the labour organisations, etc - were driving the country towards secession and civil war. In short, concluded this important paragraph of the communique, the state was left powerless and made impotent. [1]

Even if the reasons for intervention given in the communique are valid, the question arises as to why the coup occured when it did. The actual seizure of power should have surprised no one, because such an intervention had been talked of at regular intervals throughout 1980, if not earlier. And so the timing is of some interest. Why the 12th of September? Why not earlier? If we look at Turkish politics during the past year or two, we find that all of these factors were equally compelling on earlier occasions.

The press, both domestic and international, has made much of the loss of life from terrorism as the trigger for the coup. [**] If it was loss of life that was bothering the generals, then they could have certainly saved many more by intervening much earlier. But they didn't.

---

* This is an edited version of the article first published in MERIP Reports, No.93, January 1981.

** For example, the Sunday Times (London) of 14th September carried the headline: "The 4241 Reasons Why Turkey's Tanks Rolled".

Perhaps the growing intransigence of the political parties was indeed a factor of importance. The collaboration between the Republican People's Party (RPP) and National Salvation Party (NSP), the Islamic party, which led to the successful motion of censure against the foreign minister and his resignation on the 5th September, was indeed a turning point. This collaboration was important because it made an early general election virtually impossible. And it was to such an election, marked by a conclusive victory for the conservative Justice Party (JP), that the dominant faction of the ruling class pinned its hope for Turkey's immediate future. Only a strong government with a sufficiently large majority in the National Assembly, would be able to carry out reforms and implement the austerity measures which had been introduced in January 1980. But of course this alliance between the RPP and the NSP seemed to make such an outcome impossible. And so this political impasse brought about by the RPP and the NSP may well have triggered the coup.

There may have been another reason for intervening, namely the fear that continuing political and therefore social and economic instability might lead to a military coup by junior officers outside hierarchical control, something like the Greek Colonel's coup of 1967. For some time there had been rumours of officers in the armed forces, loyal to the neo-fascist Nationalist Action Party (NAP) and its leader Colonel Alpaslan Turkes (pronounced Turkesh). It was said that Turkes would welcome intervention by such officers at an opportune moment. Perhaps this pervasive political instability was providing precisely such a moment. The question therefore arises, were such officers about to act when they were preempted by generals? This was said to be the scenario for the generals' intervention ten years earlier, on the 12th of March 1971. There has been a partial confirmation of this by purges and trials of officers held after the coup. So far we have no hard evidence to confirm this thesis for September 1980. There is only some vague circumstantial evidence which may now be presented.

It seems as though the constant use of the armed forces to govern the country under martial law was undermining morale. In his Victory Day address of 30th August, Kenan Evren, who was Chief of Staff, said, "We believe that martial law should be lifted as soon as possible, in order to eliminate the adverse, even though minimal, effects it has caused within the units, despite all efforts to insure that training and fighting duties should not be affected". He noted that on average martial law has been enforced in Turkey for one year in every two over the past twenty years. Such a situation, obviously, was not good for morale and discipline in the armed forces, especially as they saw that there was no substantial change for the better in terms of restoring law and order. Such a situation was bound to lead to adventurism among some military groups. Further on in his address Evren complained that "the maintenance of peace and tranquility in the country is expected of the martial law commanders alone, and the fact that they are blamed for the failure to achieve this quickly is not compatible with common sense and justice".

Obviously the continuing crisis was being partially blamed on the armed forces themselves, at least implicitly in that they were unable to stamp out terrorism. Will the armed forces establish law and order now that they are in total control? Only time will tell. But we can see from Evren's statements that there is a strong suggestion of disquiet within the armed forces, with perhaps a potential for intervention from below.

If we pursue this line a little further we find that immediately after the announcement of the seizure of power on the 12th, General Evren addressed another communique, No.6, this time to "my valiant comrades in arms". He requested them "to fulfill the orders they had received within the chain of command, with a supreme sense of discipline and patriotism". This suggests that there may have been some other source of orders, orders conflicting with those of the National Security Council (NSC), namely orders emanating from outside the hierarchy. Then there is also the fact that, of the four party leaders who were detained, the NAP leader Alpaslan Turkes was not found in his home when troops arrived there around 3 a.m. Friday morning. He may have been tipped off. By whom is still not clear. Over the next 48 hours there were some rumours in the press that he was actually in Yozgat, a stronghold of his party. He finally surrendered to the authorities in Ankara on Sunday, 14th September, at 7 a.m. [2]

There have been no reports in the Turkish press suggesting that another coup had been preempted. But that isn't surprising since censorship was immediately ordered by the generals. [*] In his press conference of 16th September Evren seemed to suggest that the problem of a divided army was there, though not immediate. "Had this intervention not been carried out the secret and treacherous forces would have infiltrated the armed forces as well, and proceeded to divide them in a few years time". Thus, for the moment, one can say that there was a hint of restlessness in the army and that is about all. There are also rumours in Ankara of impending purges in the armed forces and the navy. Should these materialise in the near future, then this thesis will have gained some strength.

THE GENERALS' AGENDA

Far more important than the immediate cause for the intervention, and the declared aims of the coup-makers, is what they intend to do. The generals are saying what the dominant factions of the Turkish ruling class, of which they are now an integral part, have been saying since the 1960's coup about the future of Turkey. Anyone who has observed the coup unfold must be struck by the fact that it was not only well prepared technically, but ideologically as well. This is in sharp contrast with 1960, when the coup-makers had really no idea what they were going to do with their newly acquired power. They were soon taken in hand by intellectuals, mainly law professors, who wrote the liberal 1961 constitution which democratised Turkish politics but soon came to be described by the right as a luxury for Turkey. The liberal intelligentsia also came to be hated by the conservative establishment, which went so far as to inflict the indignity of torture upon it.

The coup of 1971 was better prepared. But by that time the senior commanders were an integral part of the ruling structure. Thus the fact that the commanders were well prepared in the ideological sense is not surprising. The integration of the top echelons of the armed forces into what may be conveniently described as a Turkish industrial-military complex had taken place in the 1960s and is in the process of being further consolidated with the establishment of an arms industry. That is likely to be one of the vital features of this coup: Evren constantly talks about the vital necessity of establishing a Turkish arms industry

---

* There have not been any reports in the American press either. But the English press, for example, The <u>Observer</u> of 14th September, reported that "it is widely believed that Evren led the coup to preempt a possible <u>putsch</u> by adventurous colonels".

so as not to be dependent on the outside. The implications of this for the society as a whole, and for its economy are likely to be tremendous. Turkey is not only to be an industrial power but also one that could supply arms to the region, rather like Israel does to countries of Latin America. It could do so only in collaboration with foreign enterprises, and when the Lockheed scandal broke in 1976 the speculation was that Lockheed was preparing the ground for such collaboration by bribing influential people in the Turkish establishment.

One might say that since 1960, once the coup had been taken over by the commanders, after the 13 junior members of the ruling junta had been exiled, there came into being a kind of "collective Bonapartism" in Turkish politics. The commanders began to act as mediators in civil society. Following in that tradition, General Evren and the NSC declared that they would continue to pursue the economic policies of the toppled Demirel government, something that government itself had been unable to do.

Continuity of economic policy has also been provided through the person of Turgut Ozal, who under Demirel had been undersecretary to the prime minister and head of the State Planning Organisation. The press was in the habit of describing him as the "shadow prime minister" or the "_de facto_ prime minister" wherever financial and economic policy was concerned. He was the principal negotiator with international agencies like the International Monetary Fund, the World Bank, and the Common Market. He is credited as the architect of Demirel's austerity measures of January 1980, whose fundamental aim was to establish in Turkey a free market economy, an economy which, it was hoped, would be able to find a place in the world capitalist market. These measures, among other things, ended state subsidies to the smaller sectors of industry which were dependent on such aid and essentially catered to the home market. The goal is to build up that sector of the economy which would export and be competitive on the world market.

This phenomenon is, of course, very common in the 1970s. We have seen it in operation in Chile after 1973, in Sadat's Egypt, in Mrs. Gandhi's India, and in Begin's Israel. It is a policy which has been applied not only in the underdeveloped world, but even in Thatcher's Britain, a nation in which a developed economy is in rapid decline, and seems headed for underdevelopment.

The aim of such a policy in Turkey's case is to consolidate economic power in the largest corporations or holdings, thereby making them strong and capable of competing in the world market, and more narrowly in the European Common Market. Such a policy has certain political conditions which have to be met before it can be fulfilled. Turgut Ozal's principal complaint throughout 1980, until the coup of 12th September, was that the political climate did not exist for the proper implementation of the austerity measures. On 6th September, the day after the fall of Foreign Minister Erkmen, Ozal delivered a speech which was an obvious comment on the event. Incidentally, at that point he wasn't a politician but a bureaucrat. Nevertheless, he felt competent to make political statements of this kind. It was broadcast on the radio, published in the press, and therefore received considerable coverage throughout the country. Ozal noted that "some people are again bent on placing obstacles in the way of Turkey's improving economic situation, and if these people are successful in their activity, may God preserve us, a most serious economic calamity will befall Turkey". In August, just a month before the coup, he said that he could turn the economic situation around in four years. But it was vital to implement the economic measures seriously, and that required economic stability.

Men like Ozal had placed all their hopes in an early general election and a Justice Party victory. Once early elections were no longer on the agenda, only military intervention could break this political deadlock, and implement by force the measures necessary for a functioning free-market economy.

POLITICAL REQUIREMENTS OF THE FREE MARKET

What are these political measures which will have to be implemented if such an economy can function? First of all, a free market economy will require a political consensus between the political parties. In the situation that existed before the coup, that was impossible. The small parties with perhaps 20 members of parliament, were each able to call the tune because the two major parties were so evenly balanced. They were the coalition makers: NSP leader Necmettin Erbakan could always boast that the held the key to the stability of any government. Such a situation, which made for constant political instability, had to be terminated and that meant the elimination of the parties of the right: the neo-fascist NAP and the Islamist NSP.

In the case of the NAP that may simply mean removing its leadership. It would not be difficult to establish that the party stood in violation of the constitution and therefore ought to be banned. The political space that it occupied can quite easily be filled by the Justice Party itself. In fact, the Justice Party already has a rightwing which sympathises and quite strongly with the Nationalist Action Party; thus the close collaboration between these parties after 1973.

The Islamist NSP has been declining in the country, judging by the results of the 1977 general election and the Senate elections of 1979. It is declining because its supporters have become demoralised and disillusioned because they have come to realise that it cannot deliver the goods. The party's appeal to Islamic nationalism and national capitalism, and its attacks on the multinationals, the Common Market and their Turkish allies are not convincing, particularly as the monopolies in Turkey grow stronger each year after year.

Erbakan's promise of calling in the oil-rich Islamic countries to help finance Turkey's development has also not been realised. His "failure" must surely have made many of his followers cynical of "Islamic nationalism". Thus the closure of the NSP for violating the Kemalist principle of secularism will create no major crisis in Turkey. Many of its voters may stop voting but many will return to the fold of the Justice Party.

General Evren spoke of the "Islamic revival" and the growing threat to the Kemalist republic as a reason for the military's intervention. Yet it is impossible for anyone who knows Turkey at all to take this threat seriously, as do many foreign journalists. They hint that there may be the making of another Iran in Turkey; that is surely what is implied when Konya is described as Turkey's Qum! [3] But Turkey's historical development is very different from that of Iran, and Necmettin Erbakan is no Ayatollah Khomeini. It would seem that Turkeys "Islamic revival" has already reached its peak and is decline, despite the claims of Evren and foreign journalists.

Side by side with the elimination of the small parties of the right, the new regime will also eliminate the left. The left is not and has never been a threat to the state. At the time of the coup there were five parties of the left - the Workers' Party, the Socialist Workers' Party, the Workingman's Party, the Socialist Revolution Party, and the "Maoist" Peasants' and Workers' Party - and a radical social democratic party, the National Unity Party. Despite their lip service to left unity they never came together to form a common front against the threat from the right. They lacked electoral strength and did not have a single member in Parliament. Because of their ineffectiveness, many of their supporters voted for the Republican People's Party and some even took refuge within it. There was never more than a hardcore of about twelve socialist among the RPP's parliamentary group - some said to belong to the proscribed Communist Party of Turkey - and the leftwing of the RPP was never larger than about thirty. But that was enough for Turkey's neo-McCarthyites to claim that the party was being manipulated by communists who were even penetrating the state structure while the party was in power.

Outside this legal left, there were numerous illegal factions, the most notorious being _Dev Yol_ (Revolutionary Road) and _Dev Sol_ (Revolutionary Left). Claiming to be revolutionary and indulging in terrorism in order to further their cause, they never heeded those who warned that their activity, far from furthering the cause of revolution, was merely hastening the arrival of fascism and the destruction of the left. Not one party of the left supported their tactics of terrorism, and some even denounced them as the work of _agents provocateurs_, as they probably were in many cases.

If the left in Turkey has been so inconsequential why has the state devoted so much attention to it? The answer seems to be that despite its weakness the left raised questions about the workings of the state and government which politicised people who had no leftwing leanings. Serious questions about development and social justice, Turkey's relations with NATO and the United States, the Common Market, the third world, _et cetera_ were inconvenient for an establishment that preferred consensus of the type that existed up to the early 1960s. The impact of the left was enough to force one of the major parties - the RPP - to take note of the trend and adopt a "left-of-centre" political role. Later this party under Ecevit's leadership adopted the mantle of social democracy, hoping to establish in Turkey a "capitalism with a human face" and provide an alternative to the robber-baron variety proposed by Demirel's JP. But instead of merely tinkering with and reforming the system, Ecevit spoke of the need to transform it. Moreover, he emphasised the need to maintain the liberal democratic framework which guaranteed all the freedoms, and even proposed legalising the Communist Party as a part of the democratising process.

All this talk of democracy was too much for the right, and it seems as though the ruling junta will eliminate the social democratic element from the RPP. The generals and their advisors are convinced that not even social democracy can be tolerated within the new political order they intend to establish. Thus the largest number of members of Parliament arrested after the coup were Republicans representing the left and social democratic wing of the party. At the same time the junta seems to be promoting a new leadership under Professor Turhan Gunes (pronounced Gunesh), who was in Menderes' Democrat Party until 1956 when he joined the Freedom Party. After the coup of 1960 he joined the RPP and remained in its liberal centre; he later became a focus of opposition to Ecevit. He may be able to lead an RPP purged of its left, a party very different in rhetoric from the RPP of the 1970s. Only then will Turkey have a two-party system of the type familiar to the West, one in which the voter cannot tell the difference between its parties, and political participation and enthusiasm give way to indifference. In such a system opposition will be restricted to details because the two dominant parties will be in full agreement on fundamentals.

The junta will implement and institutionalise the new order with a new constitution which excludes the liberal democratic freedoms guaranteed by the 1961 constitution and already curtailed after the military intervention of March 12, 1971. The right has spoken of the need for constitutional reform for some time and in May a plan, modelled on de Gaulle's 1958 constitution, was unveiled. Ranged behind this scheme,

> "are powerful members of the right-centre Turkish establishment, including distinguished academics and leading businessmen...
>
> The main elements of the plan are a strengthened executive and increased power for the president, who would be elected by the nation and not by parliament, and in certain clearly defined circumstances would have the power to dissolve parliament and call a general election. The senate would be replaced by an advisory body made up in part of 'life peers'...More important, a new system of proportional representation, weighted in favour of the major parties would make it harder for splinter groups to gain seats.
>
> The judiciary would be reconstituted along French lines. In extremis the president would also have the right to proclaim a state of emergency and rule by decree, though parliament would not be dissolved..." (Economist, 24th May, 1980)

The new constitution will be bolstered by a new political parties and election law which will merely spell out in detail the provisions of the constitution.

THE WORKING CLASS

The new regime in Turkey is also determined to establish tight control over the labor force and that will mean eliminating the only militant and political confederations of unions, DISK (Turkish acronym for the Confederation of Revolutionary Workers' Unions). One of the first acts of the generals after they had seized power was to announce the closure of DISK - along with the ineffectual Confederation of Nationalist Workers' Unions (MISK, a creation of the Nationalist Action Party) and Hak-Is (pronounced Ish - the Islamist union supported by the National Salvation Party). The closure of these three organisations suggests "even-handedness" towards the left and the right. But that is an illusion, for MISK and Hak-Is are token organisations with virtually no following in Turkey's working class. DISK, on the other hand, is the only challenge to the exploitation of the employers. Thus Turk-Is, the Trade Unions Confederation of Turkey, a body organised on the American model in which unions stay out of politics, is now the sole representative of the working class in Turkey. Turk-Is can hardly fill the political vacuum left by DISK, and its attempt to do so is likely to lead in time to greater militancy among the working class, the reverse of what the new regime wants to achieve.

There will of course be ideological adjustments in order to rationalise all the changes the generals want to make. The period of competing ideologies which began in 1960 has given way to the monolith which the generals describe as Kemalism. The original Kemalism of the 1930s was based on six principles: Republicanism, Nationalism, Populism, Etatism, Secularism, and Revolutionism/Reformism. The new regime is likely to emphasise all these principles to some extent or other except etatism, which is in violation of the "free market" principle. But the principle aim of Kemalism was - and will be once more - to deny the existence of class struggle and profess the common interest of all classes in Turkey, with the state mediating any differences between them. This approach may be described as the bureaucratisation of class conflict. We have already seen this method at work when 51,000 striking workers - 47,000 were DISK members - were negotiating collective bargaining contracts and were ordered back to work. National interest, claimed the junta, required that the workers get back to work while negotiations were reopened. As an inducement to negotiate workers were given a 70 per cent wage increase while negotiations were being conducted, to be deducted if the wage increase negotiated turned out to be less than 70 per cent. Seventy per cent may seem a hefty increase, though not in a society which has witnessed inflation rates of 80 per cent to 100 per cent over the last few years, so that 70 per cent has already been eaten up by inflation. But the increase is more than the employers would have conceded, and as such the directive from the junta seems even-handed. Hereafter the junta will exercise paternalism over the workers until conditions have been created for Turk-Is to take over the task.

The restoration of the ideological monolith will be marked by the end of debate in the schools and universities and in Turkish intellectual and cultural life in general. This trend is already evident from a cursory glance at the Turkish press since 12th September. Political columnists who looked at the Turkish scene critically are reduced to writing about Istanbul restaurants and football matches. Academics who refuse to toe the line are likely to be purged. This is regression in the true sense, for it takes Turkey back to the period before 27th May, 1960, when another military intervention - also in the name of Kemalism - introduced liberalism and the social state in Turkey.

THE GENESIS OF THE CRISIS

The military intervention of 12th September cannot be understood if located only in the general crisis Turkey has been undergoing during the last few years. A great deal has been written about this crisis, highlighted by "terrorism of the left and right", massive unemployment of 20 per cent (though that is surely a conservative figure), the country's bankruptcy with a foreign debt of around $17 billion, industry which works far below capacity and exports which can just pay Turkey's oil bill. What this adds up to but also conceals is the crisis of capitalism in Turkey whose origins are to be found not in OPEC price increases but in the history of Turkey's capitalist transformation.

Until the turn of this century, until the Young Turk revolution of

1908, Turkey was a society without a native bourgeoisie. The leaders of this revolution were quick to realise that westernisation meant not merely aping the West by adopting superficial reforms, but adopting its economic structure, namely capitalism. They concluded that capitalism required a capitalist class, and during the brief decade during which they were in power they went about creating a bourgeoisie. [4] They hoped that such a class would go on to create capitalism in Turkey.

The Turkish Republic established in 1923, continued the same policy, but found that the new class needed state support to grow and mature. The policy of _etatism_, adopted in the 1930s, was intended to accomplish that. During the next decade the Turkish bourgeoisie prospered, especially during Second World War when fortunes were made and capital accumulated for future investment. By 1945 the bourgeoisie had become sufficiently confident to challenge the bureaucratic state which had nurtured it. It demanded the end of bureaucratic restraints and multiparty politics in place of the monopoly state. [*] Its demands were timely because the monoparty fascist regimes had just been defeated by the liberal democracies. The start of the Cold War aided this process, and with the proclamation of the Truman Doctrine and the Marshall Plan the Turkish bourgeoisie came to believe that American capital would take over from the Turkish state the role of developing capitalism in Turkey. [5]

The period from 1945 to 1960 was one in which commercial capitalism made great advances, and during which agriculture was commercialised so as to provide food for a starving post-war Europe. That, after all, was the principal reason for extending Marshall Plan funds to Turkey. Throughout this period - at least until 1958 - the lira was overvalued at TL2.8 to the US dollar. That made foreign imports ridiculously cheap and extremely profitable as they could be sold at a four to five hundred per cent markup. But such a policy discouraged industry, which could hardly compete against the cheap imports. A similar situation exists in Britain today with its overvalued pound, artificially buoyed up by North Sea oil. In Turkey there was no such cushion and this tremendously inflationary policy strengthened a profiteering bourgeoisie. It was ruinous to those on fixed wages and salaries, that is to say, workers, state officials and soldiers. Businessmen who had government connections and were able to acquire import licenses prospered and accumulated vast sums. Generally speaking the country seemed to be booming, having been opened up to foreign capital and private enterprise after a generation under the stifling rule of the monoparty state.

And yet this policy of economic growth without proper foundations was creating a situation in which the bottom could fall out anytime. By the mid-1950s the bourgeoisie was at a loss as to what to do. The ruling Democrat Party split, with its liberal faction breaking away to form the Freedom Party. The faction led by Prime Minister Menderes asked for time and patience so that his policies could work. The Republicans in opposition called for restraints and control. In the end it was the IMF which intervened and forced the Democrats to accept its "its stabilisation programme", which among other things devalued the lira, making it 9.5 to the dollar. The IMF's intervention in August 1958 did not resolve the political crisis marked by a divided bourgeoisie. Only a force outside the bourgeoisie, namely the armed forces, could resolve this crisis by intervening and laying down new ground rules. The armed forces did so on

---

* "Monoparty" is used in preference to "single party" in order to emphasise the coalescing of party and state.

27th May, 1960, beginning a tradition of what may be described as "collective Bonapartism". Initially the bourgeoisie welcomed the intervention, grateful for the opportunity to start afresh.

In the sixties a fresh start was made. Menderes' free-wheeling policies made way for planning in which the state would play an important role under the umbrella of a "mixed economy". The experience of the 1950s had shown that private and foreign capital would not invest in ventures in which profits were low or long-term. The state would therefore have to step in and invest public money in order to create an industrial infrastructure, indirectly subsidising the bourgeoisie. That was the meaning of mixed economy. The inspiration for such planning came not from the Soviet Union, as it had in the 1930s, but from India which was touted as a model for development that was an alternative to socialism.

The 1960s was the decade of industrial development, and the economy grew at a rate of almost 7 per cent *per annum* in the years after 1963, when the First Five-Year plan was introduced. In these years large enterprises began to emerge, often in collaboration with foreign capital. They produced consumer goods for the Turkish market rather than for export, and their rationale for this was "import-substitution". In fact a captive protected market in which shoddy goods could be sold was far more profitable than a foreign market in which Turkish goods competed with those of advanced industrial states. But import-substitution is not an adequate description of Coca Cola and a variety of other western soft drinks. Instead of substituting for an import, they displaced an already existing soft-drink industry established at the local level throughout Anatolia. The same process took place in the western world, most dramatically in the beer industry which was locally organised until the conglomerates took over.

Thus in Turkey industrialisation undermined local small-scale industry and the political implication of this trend became evident by the late 1960s. This process of erosion of the small producer is graphically drawn in the figures of the State Institute of Statistics. In 1963, of 160,000 enterprises employing 10 workers or more, 178 produced 50 per cent of total output. The major companies controlled credit and the money market by setting up their own banks to finance their enterprises. But this starved the small producer of capital, leading to an increasing number of bankruptcies in the 1960s and 1970s. In the course of the year between 1969-70 the number of recorded bankruptcies increased by 20 per cent. [6] One would expect this trend to have discouraged the establishment of small enterprise, but the number of small enterprises is in fact increasing. This is clear from the figures for 1963 and 1967, and may be explained by the fact that Turkish workers abroad try to set themselves up, at least initially, as small independent producers. Some of them become successful if they are fortunate enough to choose a line which services a major industry, for example garages servicing Turkey's automobile industry. Most of them go under.

POLITICAL CONSEQUENCES OF INDUSTRIALISATION

This process has had serious political implications. For one thing, such people were alienated in the late 1960s from the principal party of the right, the Justice Party, which they correctly saw as the party of the monopolies and foreign capital. Their associations appealed to Suleyman Demirel to act on their behalf and check this process. When the failed to respond they reacted by supporting splinter groups within the JP or other parties in the political spectrum. This accounts for the formation of the Democratic Party, the Islamist National Order Party (later NSP), the Reliance Party (later Republican Reliance) in these years, and the emergence of the neo-fascist Nationalist Action Party led by Colonel Turkes from the anachronistic Republican Peasants' Nation Party. With the fragmentation of the right, the JP found it more and more difficult to retain its hold even in parliament.

A similar process was at work in the countryside, where the small holder was losing his land to the capitalist farmer who used a tractor and modern methods of cultivation. For example, the number of tractors in Anatolia had increased from about a thousand in 1945 to about one hundred thousand in 1970, while the area cultivated by tractor had risen from about one per cent in 1945 to around 30 per cent in 1969. That led to migration from the countryside to the towns and cities and the creation of shanty towns; it also enabled Turkey to export labour to the booming industries of the West, especially Germany. But the extension of mechanised agriculture also threatened landlords who lived off rent, for the capitalist farmers began to clamor for land reform, supported by the industrial sector and the two major political parties. Such landlords soon abandoned the JP and found refuge in the Democratic and Reliance Parties, further weakening the dominant party of the right.

If developments of the 1960s brought about the fragmentation of the right in Turkey, they also had the effect of creating an influential left for the first time in Turkey's history. The liberal 1961 constitution permitted ideological debate outside the Kemalist framework and as a result socialist and social democratic political clubs made their appearance and began to flourish. In 1961 trade unionists and some socialist intellectuals founded the Workers' Party of Turkey (WPT). Though never an electoral threat, it made a profound impression on the political scene through its criticisms of internal and foreign policies of the Demirel government, adding a totally new dimension to Turkish politics. In the 1965 general election the Workers' Party won 15 seats despite great odds, in a campaign marked by violence and intimidation against it. The RPP leadership took note and decided to meet this challenge from the left by adopting some of its radicalism and taking the RPP to an ambiguous position left-of-centre. That decision was perhaps the first step in the rise of Bulent Ecevit who, by 1972, became the leader of the social democratic alternative to "robber baron" capitalism.

The industrial bourgeoisie of the 1960s also created its own nemesis in the form of the working class. By the end of the sixties the industrial work force was about three million strong out of a total work force of 12 million. Of the three million about a million were unionised. Since 1963 the working class had been given the right to bargain collectively and to strike on behalf of its economic demands. But despite their increasing power, the workers were unable to obtain wage increases in keeping with rising productivity. Between 1964 and 1970 productivity increased from the base figure of 100 to 125.8, while real wages increased from 100 to 119.0 in the public sector and 110.8 in the private. [7] And yet the reason that the employer gave for keeping down wages low productivity!

In this decade the working class became more politicised and militant. Represented and educated to some extent by the Workers' Party, it was no longer willing to stay outside politics, fighting only for economic demands; it was clear the economics could not be separated from politics. As a result, there grew a movement within *Turk-Is* which demanded that the confederation begin to play a political role. When the leadership resisted this demand, a group of unions broke away to form *DISK* in 1967. Within a short time of a third of *Turk-Is* membership had defected to *DISK*, which threatened to become the dominant confederation. That did not augur well for the employers who preferred to deal with collaborationist *Turk-Is*, which invariably co-operated with management and whose leadership was closely tied to the JP despite its "apolitical" posture.

In June 1970, the JP government attempted to halt *DISK*'s growth by amending the Unions Law. According to the amendment, unions in a given industry could form a federation only if at least one third of the insured workers in that industry were members of the unions wanting to form the federation. *DISK* opposed this change, claiming that it deprived the workers of any choice since they would be forced to join the majority union, i.e. *Turk-Is*. To demostrate its support *DISK* appealed to the working class to protest against the amendments of the Unions Law. The support of the workers - both *Turk-Is* and *DISK* members - was overwhelming and on 15-16th June the industrial region around Istanbul and Kocaeli was paralysed by what amounted to a general strike. [*] The government described this affair as a "dress rehearsal for a revolution" and declared martial law in the industrial provinces. It was clear that a civilian government could no longer "discipline" the workers or establish law and order through normal means. Even before 15-16th June there had been talk of imposing the "Yahya Khan formula" modeled on Pakistan's military dictatorship, and nine months later the commanders asked the government to step down and took control. That was the coup by memorandum of 12th March, 1971.

TURKISH POLITICS AFTER 1971

These internal developments in Turkey took place in the context of the world capitalist order of which Turkey was and is an integral part, and whose health and disorders have a direct bearing on Turkey's condition. Just as Turkey benefitted from the western boom of the early 1960s, it suffered the consequences of the crisis of 1968-69 and the recessions of today. Turkey's contemporary economic problems are blamed on OPEC oil price existing problems rather than create them. By the 1970s Turkey's industrialisation had created contradictions which have yet to be resolved.

As always, the first task of the military regime set up in 1971 was to restore "stability" and that meant crushing the left because it was

---

* It is worth noting the spontaneous character of the workers anger and frustration, according to *Le Monde*'s (Weekly Section, 24th June, 1970): "The tens of thousands of demonstrators who rampaged the streets of Istanbul Monday night and Tuesday morning, sweeping through police cordons and even a few barricades that troops had set up with armoured vehicles, were not acting on orders from the *DISK* leadership.

"The rank and file had already taken matters into their own hands by forming committees in the factories "for the defence of the constitution". The demonstrations themselves were largely spontaneous. In fact, *DISK*'s leaders denied responsibility for the violence, which apparently frightened them, and in their appeals for calm blamed it on "provacateurs".

held responsible for politicising the country, especially the workers. The Workers' Party was dissolved, as were other organisations influential on the left. Freedoms granted by the constitution were curtailed, and influential public figures of the left - especially journalists and writers - were tortured and imprisoned. Between March 1971 and October 1973, the left took a beating from which it has yet to recover.

For industrialist and businessmen these were years of plenty, when the monopolies consolidated their gains at the expense of small producers. According to Ahmet Aker, the Turkish economy of the 12th March period saw two important developments: a greater dependence on western capitalism and an unexpected degree of monopolisation. It is not that dependence was anything new for Turkey, but its growth was used to speed up the process of monopoly. [8] These were also the years when real wages declined drastically, for the workers were left defenseless and unable to back their demands because of martial law. This growing confidence of the monopolies is illustrated by a full page advertisement in the press of 2nd August, 1971, announcing the founding of the Association of Turkish Industrialists and Businessmen, better known by its Turkish acronym _TUSIAD_. It started with a membership of 114 of the largest industrialists and businessmen including Turkey's press lords who own the largest circulation newspapers in the country: _Hurriyet_, _Milliyet_, and _Tercuman_. The principal aim of this association was to act as the lobby of large enterprise in Turkey. As such it maintained contacts with the parties supposedly representing small enterprise, namely NAP and NSP. _TUSIAD_ soon became a political factor to be reckoned with, though, as we shall see, still not the controlling political factor.

Side by side with these developments, there was another phenomenon taking place which created great hope and expectations in a period of darkness. I refer to the emergence of the social democratic alternative under the leadership of Ecevit, who took control of the RPP from the conservatives in 1972. The Republicans were now able to develop a broader social base by winning some support among workers and peasants, the petty bourgeoisie and the intelligentsia. For the first time the voters were given what seemed like a genuine alternative. In the long election campaign which began in the spring of 1973. Ecevit promised to defend the small shopkeeper from the predatory advance of the "supermarket" and the department stores. He promised to create a new sector, the "People's Sector", in which the small producer could invest without being swallowed up by the monopolies. _TUSIAD_ had meanwhile also announced "companies open to the people" in which the same class could invest its capital, and presumably maintain its social and economic position. The working class and the left, having no alternative, decided to support the RPP. A month before the general election _DISK_ appealed to "workers, peasants, artisans, petty officials, and all citizens who were underprivileged to vote for the RPP as the sole party which defended constitutional freedom and democratic rights". The popular response to Ecevit's campaign seemed overwhelming. But Ecevit was disappointed when his party won only 33.3 per cent of the vote and 185 out of 450 seats. Demirel, despite his dismissal by the armed forces and the charges of corruption against his family, received 29.82 per cent of the vote and 149 seats.

The Justice Party had been in disarray even before March 1971. In most countries where a bourgeois-democratic tradition was firmly established, where in fact a strong and united bourgeoisie was in existence, a party leader in Demirel's position would have resigned as a matter of course. Then the party would have had the opportunity to regroup and adopt a new image under new leadership. But in Turkey - as in other societies (India, for example) where the bourgeoisie is weak and underdeveloped - a political party comes to be associated with the leader rather than ideas or a platform. [*] Demirel followed that tradition and did not step down. It is possible that had he done so that vote of the right would have coalesced once more in the JP. As it was, the five parties of the right won 62.16 per cent of the vote and 258 seats. But the Democratic Party and the NSP were unwilling to serve under Demirel, and without their support there could be no coalition of the right.

Business circles were unhappy with the results. They would have preferred a strong, stable government of the right, capable of maintaining the high level of economic growth that took place under martial law between March 1971 and August 1973. Feyyaz Berker, TUSIAD's president, said as much on 15th October: "The election results bring to mind the question whether coalition governments will be able to maintain today's rate of development. It was my hope that the results would be such as to promote an even higher rate of development. However it seems very difficult to attain the goal with coalitions". [9]

SOCIAL DEMOCRACY AND THE BOURGEOISIE

This may have been the general feeling in TUSIAD, but TUSIAD's most advanced group - Istanbul group - seemed willing to give social democracy under Ecevit a try. After all, it could prove to be a better guarantor of social peace than a rightist government, just as it often was in Europe. Ertugrul Soysal, president of the Istanbul Chamber of Industry, told the press: "It is still too early for me to comment on the subject of a coalition...But the [election] result is quite normal. Turkey does not need fascism at this stage of development and industrialisation. If it hates communism it can get by with social democracy during this stage..." [10] Does that imply that Turkey might need fascism at another stage of development? Soysal's was a minority voice and the business circles tended to oppose any government led by Ecevit. That led to a political crisis which lasted over three months, reflecting the deep-seated divisions within the bourgeoisie.

During those months a number of different coalition formulas were tried, but all failed until the unlikely coalition between secular RPP and the Islamist NSP was formed in January 1974. It is worth noting that the political crisis coincided with the recession in Europe, which had a predictable impact on Turkey. As the European economies slowed down, the governments told Ankara to slow down export of labour. Already in November 1973, three million (18 per cent of the workforce) were unemployed, with 165,000 added to this figure each year. A million workers were waiting to go to Europe. Not only did Europe's recession affect unemployment, but it had a disastrous impact on Turkey's balance of trade, since workers' remittances had been responsible for closing the gap between imports and exports. These two factors, unemployment and balance of trade, soon led to an ever-growing foreign debt, which became the major issue of the 1970s.

Despite this critical situation, the bourgeoisie could not agree on a political compromise. Some wanted a coalition of the right, others an RPP-JP coalition, but Demirel refused to serve under Ecevit (who had the most seats), and Ecevit refused to serve in a coalition led by a non-

---

* It must be noted that Ismet Inonu continued to lead the RPP even after three electoral defeats in the 1950s.

party prime minister. An RPP-JP coalition enjoyed great support in TUSIAD, and even Turk-Is, for the business community believed that the Justice Party would be able to restrain Republican radicalism and provide strong government at the same time. Yet these people lacked the strength and unity to bring about such a solution.

A sign of this division was Ertugrul Soysal's resignation as president of the Istanbul Chamber of Industry on 8th December. His board supported the RPP-JP formula; he opposed it, believing that the business community ought to support Ecevit as Ecevit alone could guarantee social peace since his strength lay in urban areas and among the workers. "I am convinced that the RPP's majority in our country will bring a policy which brings smiles to our faces and save Turkey from social explosions". [11] Soysal did not believe that Ecevit would become an extremist in power and take his party further to the left. In his view, the experience of Western Europe proved that answers to the problems of industrial society, which Turkey was becoming, were provided by social democratic regimes of the type Ecevit wanted to establish. But the majority in his board and in TUSIAD disagreed. They were alarmed by the RPP's rhetoric against monopoly capitalism as expressed in its electoral programme, and its promise to salvage the petty-bourgeoisie of small producers and shopkeepers. [*]

The political drama was played against the backdrop of an economic situation, daily aggravated by another problem outside Turkey's control, namely inflation. The inflation rate in Turkey in the years 1961-71 averaged 7.4 per cent, the highest in Europe after Iceland. By 1973 it had risen to 9.4 per cent, the fourth highest after Greece, Iceland and Portugal. [12] Thereafter it continued to soar until it passed the 100 per cent mark in the late seventies. The oil price increase of 112 per cent made the economic situation even worse. In 1974 Turkey would have to find $800 million to pay for its oil imports, the equivalent, noted the director of the state oil company, of building two Keban dams a year. The Turkish economy simply did not have the strength to bear this burden. [13] Thus a month after he came to power Ecevit had to announce huge price increases: petroleum goods went by nearly 74 per cent, sugar 25 per cent, cement 52 per cent, state produced textiles between 20 per cent and 70 per cent, and paper 36.5 per cent. [14] Prices of all other commodities followed the trend.

The private sector's hostile wait-and-see attitude towards the government did not help matters. Its spokesmen explained that the reasons for the hostility were psychological rather than based on government measures. The statements of various ministers in both the RPP and NSP were unsympathetic to private enterprise, as was the response of certain ministries; Trade and Industry for example, to various applications. Ecevit's constant talk of creating a "People's Sector" suggested that this party was opposed to opening up the private sector to the people, precisely what Turkish capitalism needed for expansion. The more backward and conservative elements in the private sector were convinced that social democratic parties protected and supported labour rather than the employers, and that Ecevit was determined to do this in Turkey. Try as

* By 1973 TUSIAD seems to have begun considering Brazil as a model for Turkey. It published a pamphlet Bir Endustri Ulusu Doguyor: Brezilya (Brazil: An Industrial Nation is Born), translated from a publication of the Union Bank of Switzerland in 1973. Throughout the 1970s the Latin American model continued to appeal to the bourgeoisie and has been put forward by some of Turkey's only option.

he might, Ecevit could not convince the private sector that its criticism and anxiety were unjust and unfounded.In that psychological climate, the private sector refused to make investments and production began to decline, leading to greater unemployment.

The coalition was in any case unstable. Necmettin Erbakan, frightened of losing his party's identity within the RPP, was constantly obstructing government measures in order to stand out. The coalition was limping when, in July 1974, the government decided to intervene in Cyprus in order to maintain, it claimed, the status quo on the island threatened by the Samson coup. The intervention brought Ecevit great prestige, convincing him that an early general election held in the prevailing climate of euphoria would bring his party the majority he needed to rule without the restraints of a coalition. He resigned on 18th September, 1974, but soon learned that the parties of the right would not permit an early election which was bound to go against them.

If the right was united in opposition to an early election and another RPP-led coalition, it was still divided on the issue of forming a coalition of the right. The country was, therefore, forced to go without real government for over six months until 31st March, 1975, when Demirel announced the formation of the first Nationalist Front government. Again the business community pleas for either an RPP-JP coalition or an above party "national unity government" went unheard; such a government was formed by Sadi Irmak but failed to obtain a vote of confidence from the assembly on 29th November. Vehbi Koc (pronounced Koch), the doyen of Turkish capitalists, issued the warning that time was running out for Turkey, and that the Third World War on the social and economic plane had already begun in 1973. Despite the critical situation Turkey had been wasting time since October 1973 and accomplished nothing. Given the world crisis, he said, "We do not have much time to lose". [15]

TERROR FROM THE RIGHT

At this point, Demirel adopted an anti-left and anti-communist rhetoric as a strategy to unite the right. This strategy had been used in the late 1960s but became redundant after March 1971, when the left was totally crushed. There was still no organised left in 1974, and Demirel was not saying that there was. He was simply saying that he would "form a government in order to fight against the left and communism". "Me and my party are opposed to every type of left (meaning left-of-centre RPP). It is normal that the government I form should have such an identity". [16] Empty words perhaps, but dangerous nevertheless for they created a climate in which anything or anybody the right described as leftist was open to physical attacks. Rightist violence has already begun during the RPP-NSP coalition, [17] before Ecevit's political amnesty released political prisoners, later blamed by the right for the political violence in the country.

By December there was talk of forming a Nationalist Front (NF) that would isolate the Republicans as a minority. [*] But it was another four months before the government was actually formed. In the process the

* The Democrat Party had tried the same tactic in 1957 with its "Patriotic Front". The Democrat Party was dissolved following the military coup of 1960, and is considered the predecessor of the Justice Party. It should not be confused with the Democratic Party, a grouping which split from the Justice Party in early 1970s.

Democratic Party began to fall apart, virtually disappearing from the political scene by the end of the decade. The first NF coalition included the conservative JP, the Islamist NSP, the neo-fascist NAP, and the anachronistic Republican Reliance Party (RRP) composed of the extreme right who defected from the Republican People's Party when it adopted the left-of-centre position. A government composed of such contradictory elements was kept together by its fear of an early election, and its only task was to prevent that eventuality. Such a government was quite incapable of implementing any programme, let alone the kind of programme Turkey needed to cope with the world crisis, undermining the country's economic viability day by day. [18] TUSIAD, alarmed by the drift, asked the government to deal immediately with economic questions. Most of its members concluded that under present day circumstances Turkey needed a coalition that would include the RPP (i.e., an RPP-JP government). If that proved impossible the country should hold an early election. But there was also a faction within TUSIAD which described the RPP as the party which destroyed the peace of the nation, and Ecevit as the principal provocateur. [19]

For those who had supported Ecevit in the hope that he would restore pre-1971 liberalism and democracy, the NF government came as a great shock, effectively negating their vote in the 1973 election. The elements, which included workers and students, were further alienated by the NF's partisan politics directed solely against Demirel's broadly defined "left". State officials who did not support the NF were dismissed, as the parties, particularly NAP and NSP, began to fill the bureaucracy with their own supporters. This was an alarming trend, for the NF was abandoning the myth of a neutral state, one of the pillars of liberalism. Once the opposition in Turkey could be described as traitors, and the struggle as "a struggle between traitors and patriots" [*], it was perfectly legitimate for the state to support the "patriots". The "patriots" referred to those who were members of the League of Idealist Youth, the unofficial paramilitary wing of the neo-fascist NAP, whose "commandos" spread terror in the streets. That, after all, seems to have been the role assigned to the NAP in the Front, with the party being given two cabinet posts - Turkes was Deputy PM and Mustafa K. Erkovan, Minister of State - even though it had only three members of parliament. It was with good reason that the pro-NF press popularised the slogan: "Demirel in Parliament, Turkes in the Street", [20] a division of labour that continued to operate even when Demirel was out of power.

The origins of organised terrorism are to be found in this period and the responsibility for it falls squarely on the shoulders of the right. Despite the evidence, the right continued to deny its role in the terrorism of the 1970s, always blaming the left. Confronted with proof of NAP complicity, Prime Minister Demirel is said to have responded:

> "they (idealist terrorists) are our children, it does not matter if they have gone a little overboard: they are nationalist and anti-communist youths and no great harm can come from them. The NAP may be said to be more or less on our side: moreover they are the only political organisation who are not raising their voices against us. Thus to get into a conflict with them at this point would go against political sense." (Yeni Forum, 1st September, 1980, p.5)

---

* The words of Minister of Justice, Ismail Muftuoglu

Only very recently has Demirel "finally acknowledged the fact that extreme rightist terrorism has also played a role in the political violence in Turkey". Professor Aydin Yalcin's Yeni Forum describes this as "an important step forward". [*]

This is not to deny that elements of the left have also played a role in the terrorism of the 1970s. But such elements have yet to be identified by the authorities. Yeni Forum may be correct in seeing the "hand of the Soviet Union (and its satellites), Armenian, Kurdish, Greek and Palestinian elements in today's disorder, all trying to subvert Turkey from within. [21] So far all this has proved to be mere speculation. Even if it has some validity, it cannot compare in scale and intensity with the state-sanctioned terrorism of the right. The camps in which rightist militants received "commando training" were supported by the Ministry of Youth and Sports, and the Ministry of National Education. According to some estimates, perhaps exaggerated, 100,000 youths were receiving para-military training; the conservative figure of 10,000 is far too low. [22] Whatever the figure, the left could not possibly compete in either numbers or training facilities.

THE TRADE UNIONS AND THE SOCIAL DEMOCRATS

The foundations of the crisis that culminated in the military takeover of 12th September, 1980 had been firmly laid by the mid-seventies. As time passed the crisis only deepened with no solution in sight under the existing regime. The US arms embargo and Turkey's isolation in the world over Cyprus made the government seem even more impotent. Yet now that Ecevit was in opposition there was again a sense of hope and euphoria in the country that an election would bring the RPP to power, this time with a substantial majority. One aim of rightist terror was to undermine this optimism through an organised campaign of violence against the RPP and its supporters.

As a response, Ecevit persuaded the two federations, Turk-Is and DISK, to co-operate against the rising tide of fascism. One can detect a growing sense of confidence in the working class, despite the rising unemployment and the lack of political leadership from the fragmented left parties. DISK filled this gap to some extent, though its president, Kemal Turkler, recognised that the prevailing situation required that the workers support Ecevit. But what symbolised most vividly the growing sense of strength and unity in the working class was the decision to celebrate May Day in the face of a hostile regime. 1976 was the first time in 51 years that the workers of Turkey had dared to celebrate international workers' day, and its psychological impact was bound to be great. Ecevit supported the working class movement, at least rhetorical-

* The right's primary responsibility for the terrorism of the 1970s is recognised by the junta, which has taken the firmest measures against the neo-fascists and its paramilitary organisations. Turkes and his supporters are to be charged under Article 149 of the Penal Code which carries with it the death penalty for "arming people, or inciting them to rebel or massacre each other". The right has protested by clandestinely distributing leaflets accusing "the military of striking hardest against the ultra-rightwing (sic!) National (sic!) Action Party". See New York Times, 19th November, 1980. On 18th November, General Evren himself denied rightist rumours that his administration is attacking only the rightists and not acting against communists and leftists.

ly, and in turn his own campaign gathered momentum throughout 1976. Ecevit had now become Turkey's last hope against what was generally described by his followers as the growing threat of fascism, a threat Ecevit exploited with great skill while in opposition, but against which he took no action while in power.

By spring of 1977 rumours of elections were in the air and elections were finally set for 5th June. The business community wanted an early election, hoping that it would bring with it a strong government and stability. Inflation had renewed its sharp rise and at the same time the country was suffering the greatest shortage of foreign exchange in its history. Foreign banks no longer honoured Turkish cheques and imports were declining rapidly. That had a great impact on production and employment, since factories could not operate at full capacity. To remedy this foreign exchange shortage the Nationalist Front government took short-term loans at high interest bank rates and soon brought Turkey to the verge of bankruptcy. Vehbi Koc noted that,

> "this country needs a very strong government if it is to overcome today's anarchic activities, economic disorders and political troubles, if it is to rearrange our relations with America and the Common Market, and resolve our differences with the Greeks. Because of all that, the elections of June 5, 1977 have acquired great importance for Turkey today and tomorrow and for the life of our democracy."
> (Milliyet, 23 April, 1977)

Demirel is said to have agreed to a June election - four months earlier than necessary - because he feared that the Islamist NSP would campaign more effectively during the month of religious fasting (Ramazan) and the Islamic festivals, when religious consciousness would be at its highest. Thus JP votes would be lost to the NSP. But the right was most alarmed by the tremendous enthusiasm for Ecevit and the RPP, suggesting the possibility of a landslide victory for the party. DISK and four parties of the left were tacitly supporting Ecevit: Only the so-called "Maoists", calling for "revolution now," harassed RPP meetings and provoked incidents to intimidate RPP supporters, especially on voting day. On 1st May Ecevit denounced them as the "Yellow Left", who infiltrated his party's meeting and acted as provocateurs. He was now convinced, he said, that this "left" was being administered by some parties in the government (a reference to the JP and and the NAP) and that there was collaboration between the state and these bandits. "It seems," he concluded, "that they want to hold the 5th June elections under the pressure of these bandits". [23]

THE MAY DAY MASSACRE

Ecevit's press conference on May Day 1977 coincided with a mammoth DISK-organised workers' meeting to be held in Taksim, Istanbul. Its success or failure was of great political significance for the future of the movement, a fact recognised by the left and the right. The parties of the left sent messages of support; the right denounced the meeting as a provocation. The "Maoists", however, announced that they too would participate in order to denounce DISK as revisionists and collaborators with the ruling class. But they also warned workers who participated in the meeting to be wary of agents provocateurs of the Nationalist Front who would also be active. DISK marshalls, on the lookout for provocateurs, prevented incidents.

The meeting went smoothly, but as it was about to end some shots were fired, starting panic in the massive crowd. Riot police, out in force, only aggravated the panic by turning on their sirens, designed specifically to terrify crowds. Moreover, they closed off the boulevards which branch out from Taksim with their armoured cars, forcing the crowd to run into narrow streets where many people were trampled upon. The police also instructed people to lie down to escape bullets being fired at random; many of those who did so were also trampled by the panic-stricken crowd. Of the 36 deaths reported in the press of 3rd May, only two had died of bullet wounds; the rest had been trampled to death while hundreds were wounded.

The impact of this May Day massacre - Turkey's second Bloody Sunday [*] - and of violence in general on the election results is impossible to measure. But the results again did not match RPP expectations which may well have been optimistic. Instead of the 230 or 240 seats they expected, the Republicans won 213, 13 short of the majority needed to govern alone. The right seemed to be coalescing again, to with the splinter parties being wiped out. That was the fate of the Democratic Party, whose seats declined from 45 (1973) to 1, and the Republican Reliance Party's seats from 13 to 3. Even the Islamist NSP lost half its seats, which seems to make nonsense of the so-called Islamic revival in Turkey, if indeed that was the cause of the party's original appeal. It is more probable that all these parties declined because the petty bourgeoisie learned from experience that it could not halt the progress of the monopolies by supporting them. This class, still very divided, therefore, supported the RPP (which had always promised to save "the small man"), the neo-fascist NAP with its anti-capitalist rhetoric, and the JP, whose policies led to the integration of some small producers in the growing capitalist sector. But again the results were indecisive, failing to produce the strong government the country needed.

Ecevit asked to form the government. Unable to reach an agreement with the other parties, he formed the first minority government in the Republic's history. The mood of desperation in the country in this period was summed up in the words of retired colonel who told Milliyet (26th June, 1977): "If Mr. Ecevit's government does not get the vote of confidence, God help Turkey". There was no chance of the government getting the confidence of the assembly because the parties of the right were united in their determination to bring down Ecevit even before he was on his feet. The government was duly defeated on 3rd July and Demirel was asked to form the next cabinet. Again the business community proposed an RPP-JP coalition, again to no avail. Demirel then formed his second Nationalist Front on 21st July, giving eight ministries to the NSP and five to the NAP, while his party held 16. However, there was a difference between the tow NF governments. In the first, the dog (JP) had wagged the tail (NAP); in the second there was a danger that the tail might wag the dog. Twelve Justice Party deputies, alarmed by the extremist NAP-inspired racialist trend which was dividing the country into hostile groups, resigned from the party and became Independents. They denounced the government and stated that "because of the activities of the Front government, the murders taking place, and the pressures (on

---

* The first occured in 1969 in Taksim, on the occasion of demonstrations protesting a visit of the US Sixth Fleet.

Kurds and Shi'a in Southeast Anatolia), they considered it their duty to vote against it". Thus on 31st December the government fell and a week later announced his government, this time supported by the Independents as well as by two arch-conservatives Faruk Sukan (DP) and Professor Turhan Feyzioglu. It was a de facto RPP-JP coalition which diluted what little radicalism there was in the RPP and further divided it.

ECEVIT AND TURKEY'S CRISIS

Ecevit came to power with the promise of restoring "peace and unity" in the country. It became the raison d'etre of the opposition to prove him wrong. Instead of peace and tranquility being restored, political terrorism increased in intensity: 30 were killed and over 200 wounded in the first 15 days of 1978. By July the special riot police was unable to cope and gendarmes with commando training (the Blue Berets) had to be called in.

Ecevit nonetheless professed to be optimistic about the future. He told his BBC interviewer that the IMF was impressed with the austerity measures he had introduced and credits would soon be forthcoming. Asked about domestic violence, he said that had been encouraged by the two coalition governments before his but was coming under control. Under his government, he maintained, the rightists were getting frustrated because they were losing control of the state, and were therefore putting up their last resistance. This was, of course, a grave under-estimation of the Turkish right. Asked about the violence of the left, Ecevit agreed that there were violent elements among the gauchiste groups but they were reacting to the violence rather than initiating it. He neither apologised for their violence nor approved of it, and under his administration, he said they were being arrested in equal numbers. [24]

Ecevit was being too sanguine; in fact, leftist terrorist factions (with names which can best be read as acronyms) had begun to carry out a campaign of assassination against those responsible for the 1971 regime. But gangsters also used the cover of political violence to carry out their crimes. For example, four men claiming to belong to TIKKO [*] kidnapped a businessman and demanded a ransom of TL3 million. They turned out to be notorious Istanbul gangsters! [25] The base of rightist violence broadened and spread into Anatolia, directed particularly at the Shi'a who supported the RPP as the most secular party.

There was little Ecevit could do to ease Turkey's economic predicament in a period of crisis. He merely followed the usual remedies to check inflation, which placed the burden on the already over-burdened. His prestige among the workers enabled him to sign a "social contract" with Turk-Is, and member unions who agreed to restrain wage demands. Just the kind of practice on expects from a social democratic party! But it was denounced by DISK, and in a short time lost the RPP the support of many workers, support the party had come to depend on in the industrial western region of Turkey. Faced with a foreign debt of around $17 billion, he asked the planning organisation to prepare a more liberal foreign investment code which might bring in the badly needed currency, even though that aroused strong opposition within the party. He had already introduced harsh austerity measures in March but they were insufficient to appease the IMF. The IMF expected Ecevit to bring down

---

* A leftist terrorist group, the Independence Army of the Workers and Peasants of Turkey.

the inflation rate from 50 per cent in 1977 to 20 per cent in 1978 by lowering wages and consumption, reducing the speed of development, increasing taxes, and giving priority to export industries. [26] When Ecevit failed to deliver, the IMF refused to grant badly-needed credits and negotiations between the two sides continued into 1979. By December, there was strong IMF pressure to devalue the lira, which Ecevit resisted despite the fact that

> "political and economic circles in the capital are unanimous about an impending devaluation...The same circles stress that economic relations with IMF should not be based on pride, and national sensitivity but on realism and pragmatism."
> (Turkish Daily News, December 7, 1978, Ankara)

These few words sum up Ecevit's policies and the attitude of Turkey's ruling circles (and the West) towards them. Ecevit represents those elements in Turkey who continue to see the state as the arbiter of civil society rather than the agent of a specific class. His conception of "national interest" is therefore broader than that of a specific class. That is why he was accused of having "pride and national sensitivity" by those whose "national interest" required that IMF credits be made available to them, regardless of the terms. Ecevit's conception of the state's role therefore makes him somewhat unpredictable, and in the end makes the support of all classes lukewarm. If the business community opposed his policy of resisting IMF demands, most people felt let down and alienated by his acceptance of the onerous austerity programme.

Ecevit's foreign policy was equally ambivalent and unpredictable. He did not renounce NATO or go further than Demirel in cultivating the Soviet Bloc or the Third World, but his perception of the world and Turkey's place within it differed sharply from that of Demirel and the groups he spoke for. The latter see Turkey as an intermediate nation, superior to Third World countries but inferior to the West, which they want to emulate and join. This worldview explains their policies since 1945; capitalism, as laissez faire as possible at home, and unconditional commitment to the West abroad, even though that meant disdain for the Third World (including the Middle East) and Turkey's own isolation.

Ecevit and the RPP came to regard Turkey as a part of the exploited, developing, non-industrial world which was being rapidly left behind by the West and the Soviet Bloc. Turkey, one of the first nations to liberate itself from imperialism, ought therefore to play a leading role in the efforts of oppressed and exploited countries to change the world order and to close the North-South gap. Such a perception leads to the possibility - though highly improbable - that Turkey might one day choose non-alignment. More immediately, it leads to attempts to steer an independent course in foreign policy, to the liking of neither the West nor Turkey's ruling circles.

In power at the head of the weak and captive government, Ecevit's ambivalent policy pleased no one. He was forced to proclaim martial law after the Kahramanmaras massacre of December 1978 [*], though it was limited to 13 provinces and operated "within the framework of the free democratic regime". That contradiction pleased neither the commanders who claimed they had insufficient powers and demanded more, nor the masses in the 13 provinces who lived under severe repression. In gener-

---

* See MERIP Reports, No.77, May 1979

al the country became disillusioned with Ecevit, having been led to believe that his coming to power would somehow bring about the resolution of Turkey's problems. That, after all, is charisma. But charisma depends on performance and Ecevit had been able to do very little in power. Rarely has a politician's image evaporated as rapidly as Ecevit's. The voters showed their despair and disillusionment in the Senate and Assembly by-elections of October 1979. Ecevit's support declined from 41 per cent in 1977 to 29 per cent, while the Justice Party continued to consolidate the vote on the right with 47 per cent, marking the continuing decline of petty-bourgeois support for the small parties.

Ecevit resigned on 16th October and Demirel announced his minority government on 12th November. The situation had been reversed from earlier crises, when Demirel and the small parties feared an early general election. Now it was Ecevit' and NSP's turn to do so. But it is worth emphasising that neither the public nor even the JP moderates were willing to accept another Nationalist Front government which included the neo-fascist NAP, so subversive had they been of the democratic process. On the other hand, big business opposed NSP participation because the Islamists tried to obstruct the monopolisation of the economy, though with little success. Nevertheless both parties supported Demirel, the NAP unconditionally.

Demirel described himself as the leader of an "emergency government" whose task was to manage the existing crisis. He left the law and order question to the commanders, who were given a free hand and no longer restrained by Ecevit's "framework of a free democratic regime". Economic order was to be restored by "the law of the market," as as a first step the lira devalued from $47 to $70. [27] That measure was expected to raise prices by at least 50 per cent; 40 million out of Turkey's 45 million people would suffer the harsh consequences of this inflation. [28] For Demirel and Turgut Ozal, who had just become undersecretary to the prime minister's office and head of the State Planning Organisation, the measures of 25th January were last ditch measures which had to be made to work for Turkey's survival. The economy was thrown open so that the fittest (i.e., the large companies) would survive and western money would flow in, undeterred by bureaucratic regulations. Within a few months, the small and medium enterprises saw the writing on the wall. They begged Ozal not to be "the man who wrote the epitaph on the tombstone of middle and small industrialists," pointing out that "the economic policy being implemented was totally to the benefit of the holdings". [29]

Generally speaking, Demirel's economic policy was described as a gamble. [30] A gamble, however, is a matter of chance. Demirel's scheme, though full of risks, was more than that. He was trying to establish a free market economy which would require an ideological and political consensus accepted even by the workforce. That seemed a fantasy in the Turkey of 1980. Western ambassadors in Ankara were puzzled as to how Demirel would implement measures which required a very strong government. "It was being suggested that the economic formula known as the 'Chicago School Formula' after the famous American economist Friedman, which had been implemented in Pinochet's Chile and brought favourable results there, could bring similar results to Turkey". [31] Ideas such as these become accepted wisdom. Repeated *ad nauseum*, they created a climate of despair. Turkey was said to be sinking rapidly and only extraordinary measures could keep her afloat. But where was the political climate to accomplish them? If anything, the political climate became worse until finally a political impasse was reached in September. At that point, the commanders, who had shown great patience with the politicians, stepped in, convinced that politics was now too important to leave to politicians, and that they alone could implement policies that the politicians could not.

NOTES

1. Text in Milliyet, September 13, 1980.

2. Ismail Bilen, Secretary General of the illegal, pro-Soviet Communist Party of Turkey, said in an interview on September 16 that "Turkes sought refuge within the 28th Division, which is controlled by fascists. However, following negotiations with the junta, he surrendered". "Voice of the Communist Party of Turkey", 21st September, 1980, in the BBC's Summary of World Broadcasting, ME/6531/e/5, 24th September, 1980.

3. See David Barchard, "The Religious Reactionary who triggered Turkey's Army Coup", Manchester Guardian Weekly, 28th September, 1980.

4. Feroz Ahmad, "Vanguard of a Nascent Bourgeoisie: The Social and Economic Policy of the Young Turks, 1908-1918" in Osman Okyar and Halil Inalcik (eds.), Social and Economic History of Turkey, Ankara, 1980, pp.329-50.

5. This theme is developed more fully in my book, The Turkish Experiment in Democracy. See also Caglar Keyder, "The Political Economy of Turkish Democracy", New Left Review, No.115, May-June 1979.

6. Kuter Atac, Ozer Baykay and John Bridge, "The Political Implications of Recent Turkish Economic Development," unpublished paper, n.d. The authors usually quote statistics from the various Chamber of Commerce and Industry and from the government.

7. Ibid.

8. Ahmet Aker, Bagimli Tekellesme (Dependent Monopolisation) 1975.

9. Milliyet, 17th October, 1973.

10. Ibid.

11. Milliyet and Cumhuriyet, 9th October and 10th 1973; also Ahmad, Experiment, pp.331ff.

12. Manchester Guardian Weekly, 23rd February, 1974

13. Quoted in Milliyet, 23rd January, 1974. The Keban Dam is an important project on the Euphrates in Eastern Anatolia.

14. Milliyet, 25-26th February, 1974

15. Vehbi Koc, n.d. Reproduced in Milliyet 1974, Istanbul, 1975, pp.182-83.

16. Yanki, 14-20th October, 1974, p.5.

17. Yeni Ortam, 4th May, 1974, listed incidents of political violence by the right even before Ecevit came to power. Papers like Milliyet interpreted rightist violence as being designed to create disorder so as to provoke military intervention. See, for example, Milliyet and the press generally from 21st December and 23rd, following the murder of a student in Istanbul.

18. Metin Heper, "Recent Instability in Turkish Politics: End of a Monocentrist Policy?", Istanbul Journal of Turkish Studies, I/i, 1979-80, pp.102-113.

19. Cumhuriyet, 18-19th October, 1975. Earlier Koc had written to Ecevit and Demirel stating that an RPP-JP coalition was an absolute necessity to deal with Turkey's economic woes, Cumhuriyet, 28th August, 1975.

20. Ahmad, Experiment, p.347.

21. Ibid.

22. Cumhuriyet, 25th June, 1977.

23. Ibid., 2nd May, 1977.

24. BBC World Service, The World Today, May 16, 1978. Taped and transcribed by the author.

25. Hurriyet, 30th June, 1978.

26. Ibid., 18th November, 1978.

27. The Turkish Press, 25th January, 1980.

28. Yanki, 11-17th February, 1980, cover story, "What will the People do?".

29. Speech by Murteza Celikel, member of the Istanbul Chamber of Industry Board, in Cumhuriyet, 22nd August, 1980.

30. Yanki, 4-10th February, 1980, p.5 and and The Times, 22nd July, 1980, London, "Will Turkey's Economic Gamble Pay Off?".

31. Yanki, 11-17th February, 1980, p.4.

APPENDIX 1: THE TRADE UNION MOVEMENT

In the Middle East, only Egypt rivals Turkey in the size of the proletarait. In recent decades, the working class of Turkey has developed important organisational forms which made it an important political force in the country and a prime target of the new junta.

In the first decades of the Republic a strong state apparatus, *Kemalist* nationalist ideology which denied class conflict, and a paternalistic labour code taken from fascist Italy in 1937 all worked to contain the development of trade unions, strikes, and working class political activities. In the course of political liberalisation after Second World War, though, many workers' organisations emerged - 239 separate unions by 1952. Those with leftist politics were suppressed, and considerable efforts were made under allied pressures and advice to institutionalise trade unions along the lines of those in the major capitalist countries. The American Federation of Labor (in collaboration with the CIA) provided funds and advisors in this effort, which led in 1952 to the formation of the officially sanctioned Trade Unions Confederation of Turkey (*Turk-Is*), with a membership of some 150,000 workers.

Indications of the growing political force of Turkey's working class were apparent in the 1960s. The Workers' Party of Turkey (WPT) was established in 1961 by leftists formerly associated with the Republican People's Party (RPP) and with the outlawed Communist Party of Turkey. Led by trade unionists and intelligentsia, the WPT sought to establish a constituency within *Turk-Is* as well as to provide an independent voice for the working class in Turkish politics.

This had an effect on the political direction of the RPP following the latter's defeat in the 1965 general election in which the radical vote went to the WPT. To counter this trend, Bulent Ecevit, the party's general secretary, suggested staking out a "left-of-centre" position in the political field and later advanced a line of social democracy for Turkey. The labour movement had acquired teeth for the first time with the passage of legislation in 1963 legalising the right to strike. It provided for union representation on the basis of a majority vote in any workplace of three or more workers, enabling *Turk-Is* to flourish.

One aim of the 1963 legislation, to place Turkey's working class largely out of the reach of Marxist or radical political union organisers, was countered by a new, higher level of discontent spawned by the build-up of industry, with the reorganisations, speed-ups and profiteering that accompanied it. The collaborationist character of the *Turk-Is* leadership and the anti-worker policies of the state-owned industries were exposed in a series of important strikes in the mid-1960s.

In the Zonguldak coal strike of 1965, 46,000 rank and file miners struck against the bonus distribution policies of the state managers, which had been approved by the union leadership. The strike provoked military intervention which left two workers dead. The RPP joined with other parties in denouncing the strike as illegal and the work of the "communist" WPT. The workers did win their immediate demands,though, and the *Turk-Is* leadership was discredited.

A four months strike in early 1966 of 2400 glass workers in Istanbul against a national contract led to the emergence of a radical challenge to moderate leaders in the union, with the workers supporting the radicals and management the moderates. The national union and *Turk-Is* denied the strikers access to support funds contributed by workers in other industries. The strike was eventually broken, but the sharp splits and new alliances thus engendered in *Turk-Is* led directly to the formation in 1967 of DISK, which assumed an explicitly anti-capitalist stance and encouraged direct action, such as street demonstrations, to advance political as well as economic goals. DISK in its first decade was led by Kemal Turkler, an active unionist since the 1950s who had served as general secretary of the Metal Workers' Union. He was a founding member of the Workers' Party and served on its general administrative board. (He was assassinated in July 1980.)

The leftwing unionism of DISK found its strongest support among the better-paid workers in industries dominated by larger, more automated plants in the private sector, and particularly those in the Istanbul region. In some cases these are plants in which foreign capital has a major stake, but the essential feature is the structure of capital rather than its nationality. Metals, chemical, petroleum, rubber and press workers have been responsive to DISK. In food processing and textiles, where large mechanised plants coexist with small, traditional labour intensive enterprises, the extent of radical organising is not nearly so great. In furniture and leatherwork, characterised small enterprises and low pay, there is no sign of leftist unionism. Strikes of DISK workers are characteristically long in duration, encompass significant wage and other demands, are less susceptible to compromise and have a high rate of failure. Turk-Is, dominant in the public sector, has often been pampered by the government with which it collaborates. While salaries in the public sector are higher than those in the private sector, only the militancy of DISK has prevented the gap from being larger still. This is why so many workers abandoned Turk-Is and flocked to DISK.

DISK grew to over 350,000 members an over 20 unions by 1971. Justice Party legislation designed to restore the balance in favour of Turk-Is led to the mammoth workers demonstration of 15-16th June, 1970. Although controlled by martial law forces, this event had great impact on both the working class and the ruling class, demonstrating the power of the workers when mobilised in large numbers. Prime Minister Demirel told his parliamentary group:

> "There have been a number of incidents during our term of office, but this is the first time we have faced an incident of such magnitude against life and property...What happened yesterday in Istanbul was a rebellion."

The protest forced the government to drop the anti-DISK legislation, but it also had more ominous repercussions.

The crackdown against the left after the March 1971 "coup by memorandum" makes sense only against this backdrop of growing politicisation and militancy among workers as well as students. The regime closed down the Workers' Party but not DISK, depriving the trade union movement of the political leadership of a party. Until 1971, DISK had been quite close to the WPT, especially while Mehmet Ali Aybar was its chairman. When political parties of the left appeared once more after the 1973 election, their different voices served only to confuse the leaders, if not the workers, of the various unions. In these circumstances, workers voted for the RPP with its promises of democratic rights and social and economic justice.

After 1973 almost all the parties of the left - the WPT, the Socialist Revolution Party, the Workers and Peasants' Party, the RPP, even the proscribed Communist Party - had factions in DISK. The strongest were those of the WPT and the RPP. The workers did begin to regain their confidence, shattered by the 12th March regime. They mobilised against the State Security Courts and other anti-democratic features introduced after 12th March. They fought against the neo-fascists active in the Nationalist Front government (1975-77) and against the reign of terror of the right. The high point of this mobilisation was the May Day rally of 1976. The right did not permit this courageous step to go unchallenged. The next year May Day celebrations were turned into a bloodbath by provocateurs. [*]

The inability of the RPP to form a government after winning the 1977 elections led to a proposal within DISK (allegedly from Workers' Party and Communist elements) for a National Democratic Front of all progressive forces. Ecevit saw this as an issue which would enable him to win control of DISK and turn it from a socialist to a social democratic confederation. Turkler, perceiving a challenge to his leadership from DISK's leftwing, allied with Ecevit. Turkler came out of this maneuver isolated and incapable of dealing with a challenge then mounted by the RPP faction in DISK.

Abdullah Basturk, who succeeded Turkler as DISK general secretary, was a seasoned unionist who joined the movement in 1955. In 1962 he unified the unions of municipal workers into a federation called Genel-Is. Basturk supported Ecevit's left-of-centre platform in the RPP and was rewarded with a seat in parliament in 1969, where he sat until 1977. He did not break away from Turk-Is in 1967 to form DISK, but led a movement to convert Turk-Is into a social democratic confederation. This made little headway. Basturk left Turk-Is in 1975 with a well organised union and a following of 130,000 workers. He joined DISK the following year to challenge and defeat Turkler as president of the confederation.

Basturk realised that to tie DISK to the RPP's coattails would risk losing the support of its most political and militant workers, who continued to be influential. DISK thus refused to sign a "social contract" signed by Turk-Is and Ecevit's government in August 1978, restraining wage demands. Basturk would not have been able to carry the rank and file with him in such an agreement in any case. Ecevit, under military pressure, banned May Day celebrations in Istanbul in 1979, which lost him much working class support. DISK refused to go along with the government's decision, and a 28 hour curfew had to be ordered to restrain the workers.

Turkey's working class seemed trapped in a dilemma; political castration under Ecevit, or political struggle and repression under Demirel. The choice, as it turned out, was not the workers'. With Ecevit's resignation in November 1979, they faced Demirel's attempts to tame them. The most dramatic confrontation occured in Izmir in February 1980.

Taris (pronounced Tarish - acronym of the Producers' Sales Cooperative Union) employed thousands of workers in state enterprise spinning mills and other factories. DISK was the dominant union until the Nationalist Front came to power in January 1975. Demirel appointed as Taris's general manager Orhan Daut, a former Justice Party member of parliament and general secretary of the Union of Chambers of Trade, Industry and Stock Exchange. Daut set out to replace DISK with MISK, the labour front of the neo-fascists, by systematic coercion. Daut quickly won notoriety as a "commando retainer," using them to intimidate workers.

---

* May Day was celebrated again in 1978, while Ecevit was prime minister. That was the last time; martial law authorities refused permit May Day activities in 1979 and 1980.

After Ecevit won the election of 1977 he did replace Daut but made no attempt to break MISK's control of Taris. MISK nevertheless declined rapidly during Ecevit's 22 months in office. When Demirel returned to power he was determined to reverse the situation in Taris again. This time there was resistance from the workers, supported by thousands rebelling against the harsh austerity measures introduced on 24th January, 1980. There were armed clashes and pitched battles, and troops had to be brought in before the workers finally abandoned their struggle.

Through 1980 Turk-Is continued to work with the government in power. DISK leadership, veering towards social democracy, found itself without political patronage. Once more it was forced to adopt an independent and militant position. Its leaders had to court arrest by defying martial law orders concerning May Day or risk losing support of the rank and file. This summer, DISK workers went out on strike as collective bargaining contracts came up for renewal. This was the situation when the military seized power on 12th September, 1980.

APPENDIX 2: EMPLOYMENT STATISTICS

TABLE 1

Non-Agricultural Employment Indices in Turkey (1970=100)

	Manufac-turing	Mining & Quarrying	Construc-tion	Transport, Storage, Communications
1966	74.9	97.4	70.6	92.1
1967	81.6	111.6	74.5	82.1
1968	94.2	96.1	87.7	93.3
1969	95.7	97.8	94.8	99.1
1970	100.0	100.0	100.0	100.0
1971	107.7	120.1	97.6	112.2
1972	119.8	104.4	108.0	128.6
1973	130.9	106.5	116.4	128.2
1974	124.6	120.8	129.9	148.5
1975	136.5	116.0	140.7	149.0
1976	154.1	124.4	158.3	164.1
1977	162.1	170.5	172.3	178.7

Source: International Labour Organisation, Yearbook of Labour Statistics, 1976, 1978

TABLE 2

Unemployment in Turkey (in thousands)

1966	23.5	1972	43.9
1967	26.8	1973	44.8
1968	33.0	1974	81.7
1969	39.0	1975	116.8
1970	43.8	1976	153.3
1971	44.9	1977	188.9

Source: International Labour Organisation, Yearbook of Labour Statistics, 1976, 1978

TABLE 3

Consumer Price Indices in Turkey (1970=100)

	General	Food	Clothing
1970	100.0	100.0	100.0
1971	116.3	114.0	118.4
1972	131.4	126.5	139.0
1973	153.2	151.8	174.3
1974	181.8	180.8	218.6
1975	218.3	235.1	239.8
1976	251.7	277.2	272.8

1977	323.3	362.0	358.4
1978	474.4	493.7	NA

Source: International Labour Organisation, *Yearbook of Labour Statistics*, 1976, 1978

TABLE 4

Industrial Disputes in Turkey

	Number of Disputes	Workers Involved	Working Days Lost
1968	54	5,259	176,448
1969	81	15,134	267,863
1970	112	21,150	241,226
1971	96	10,916	475,456
1972	121	13,437	628,246
1973	55	12,286	677,345
1974	105	22,922	741,397
1975	113	13,486	664,576
1976	56	7,256	395,245

Source: International Labour Organisation, *Yearbook of Labour Statistics*, 1978

TABLE 5

Wages in Turkey (in Turkish Liras per day)

	(1)	(2)	(3)	(4)	(5)	(6)
1968	28.22	27.06	27.09	29.03	33.72	28.34
1969	32.13	31.80	27.01	32.15	38.42	36.26
1970	35.32	35.72	31.39	33.72	40.41	35.45
1971	39.32	40.74	33.09	38.25	46.30	44.11
1972	43.88	45.21	35.64	41.71	52.13	38.74
1973	54.41	57.28	48.31	48.10	62.28	50.31
1974	68.26	70.92	59.60	64.51	74.17	53.93
1975	85.88	89.75	90.19	77.15	120.15	89.05
1976	115.30	126.29	112.10	105.54	117.44	99.59
1977	132.25	127.52	114.57	126.60	165.30	120.80

(1) All Non-Agricultural Sectors
(2) Manufacturing
(3) Mining & Quarrying
(4) Construction
(5) Transport, Storage, Communications
(6) Agriculture

Source:International Labour Organisation *Yearbook of Labour Statistics*, 1976, 1978

8 The state, the military and the development of capitalism in an open economy

HUSEYIN RAMAZANOGLU

The transformation of the Turkish economy from a closed to an open one is now well under way, but this transformation has not been simply a process of economic change. The restructuring of the economy brings with it major political and ideological changes, and the events which led inexorably to this transformation were very complex. In many instances the processes of transformation have been disguised as merely economic changes, but as I have been arguing throughout, this conception is misleading. The political and ideological contradictions generated by uneven capitalist development in Turkey have been at the root of the crisis of Turkey's capitalism. As Turkish capitalism has developed, the classes also developed, but in the Turkish case, the fractions of capital that dominated the economy, failed to gain effective political power. This non-correspondence between the economic and political power of the dominant fractions, highlights the acute problems of capital accumulation within Turkey.

The problems arising from this non-correspondence reached a stage in the late 1970s, where immediate solutions were urgently needed if Turkey was to continue as a credit-worthy economy and a respected member of the Western Alliance. The contradiction between the requirements of capitalist expansion and the Republic's achievements of a liberal form of parliamentary democracy appeared to have been resolved, however arbitrarily, by the military intervention in 1980, but the struggles for power which generated the contradiction still persisted. If the aims of the 1980 coup were to be successfully achieved, the abolition of the democratic system had to be swiftly and efficiently followed by action to open the economy, and entrenched monopoly/industrial and financial capital in power.

In this chapter I examine the political consequences of opening the economy, looking in particular at events between 1980 and 1983, and assess the implications of this restructuring of the economy for the further development of Turkish capitalism. I would argue that the changes already planned or implemented are transforming Turkish capitalism irrevocably, so that attempts to halt or reverse the opening of the economy are unlikely to succeed. This does not mean, however, that change has to proceed in a pre-determined direction, and the outcome of the present upheavals remains uncertain. What is certain at present is that, even with the return to civilian government, the military still hold effective power and have sufficient control of the economy to implement change, even if change does not always proceed smoothly.

General Evren and his regime quickly consolidated their power, as was shown by the 1982 referandum in which the new constitution was overwhelmingly approved by the Turkish people, making Evren President of the Turkish Republic. There were, however, still divisions within the ruling National Security Council and the power block that it represented, and these were chiefly over priorities in implementing the new measures. These cracks in the apparently united military-industrial/financial complex were indicated by bitter criticisms of the civilian arm of the military regime, which appeared within some fractions of these dominant interests. Some previously committed supporters of the regime seem to have reconsidered their position as the impact of the new stabilisation programme imposed by the IMF becomes more apparent and started to take its toll in the uncompetitive sectors of the economy. These undercurrents and the sharpening of contradiction between different fractions of the bourgeoisie sometimes reach public notice, and put the credibility of the regime to the test. In spite of these tensions, there did seem to be a general consensus among the dominant political and economic interests that Turkish capitalism should never again be restricted to development in a closed economy, and that all fractions of capital should struggle to achieve more advantageous positions in the new, open economic framework.

The efficiency and ruthlessness with which the authorities proceeded to put their programme into effect (with advice and backing from the centres of international capital) surprised everybody, not least the government itself. The degree of popular support which the military enjoyed was crucial in enabling them to implement their policies, and it seems unlikely that this was simply a happy accident. It seems more probable that the timing of the coup had been very carefully planned by the military, in association with their national and international supporters, in order to draw on the maximum possible popular support for the radical transformation of Turkish society which was intended. The coup itself came as no surprise to the Turkish people. It had long been expected, and brought immediate relief from daily violence and terror. Its initial popularity was so marked that several western observers concluded that, unlike Poles, Turks prefer stable military government to unstable democracy. The cost of political stability was the immediate ban on all political activity, and the dissolution of all political as well as socio-economic organisations and, of course, the introduction of austerity measures the like of which the Turkish people had never experienced before. To give credit where it is due, the military and its leader, General Evren, never attempted to give a false impression of their aims nor of how they intended to achieve them. They made absolutely clear to the Turkish people that the economic and political measures which they took were dependent on each other. This frank and open approach to politics had obvious appeal to the people which must be understood in order to appreciate the initial popularity of the military government. After having lived through dishonesty, deceit, lies, opportunism, political incompetence and violence as the norm of political activity, stability, austerity and the bluntness the new regime were welcomed. This feeling was evident in all strata of Turkish society and across most of the political spectrum, and given the obstacles facing them on the economic and political fronts, the military needed all the good will it could get. As the election of the 6th November, 1983 showed, this goodwill could not be taken for granted when issues affecting the direct participation of people in political processes were at stake.

MEASURES TAKEN BY THE MILITARY REGIME TO TRANSFORM TURKISH CAPITALISM

As I have suggested earlier, the military were totally committed to changing Turkish capitalism and with it Turkish society. The military, and its civilian off-shoots, will be judged on the basis of how far they could fulfil their promises and achieve their goals. Their activities can be broadly categorised as falling into two main headings: economic and political. The measures that the military regime undertook or had planned, therefore, will be looked at under these broad headings.

Economic measures

The measures imposed on Turkey by the IMF, and finally put into into effect in January 1980, were aimed at transforming the Turkish economy from one based mainly on small and inefficient businesses, and large and inefficient state economic enterprises, into a highly efficient and competitive capitalist economy based on industry and agriculture. In this section, I will outline these measures and consider their impact on Turkish economy and society.

The implementation of the 1979 IMF stabilisation package, was not possible without rapid change in previous business practices, the running of the state economic enterprises, and the established habits of consumers. Resources had to be rapidly reallocated between different sectors of the economy, in order to promote the production of goods for export and to abolish restrictions on further expansion. These changes meant that economic activities, which traditionally had been geared to a protected market would have to disappear almost overnight leaving the inefficiency of the manufacturing sector and the low quality of Turkish products exposed to foreign competition. The introduction into Turkey of cost-effective and high quality foreign goods, would necessarily undermine the position of the indigenous manufacturing sector, and increase unemployment in the industrial sector. [1]

Banking: The IMF package inevitably brought economic distress in its wake. In addition to growing unemployment, attempts to establish efficient and well organised money markets gave rise to dislocations in the financial markets. Prior to the 1980 military intervention, the major holding companies had each acquired a bank in order to meet their cash flow problems. Banks had become institutions which guaranteed to provide loans on favourable terms to their holding companies thus virtually ensuring the continued existence of these firms whether or not they operated efficiently. Although full integration was prevented by legislation and only 25 per cent of bank loans were allowed to members of the same group of companies, further legislation was desperately needed in order to separate banks from their holding companies.

When interest rates were freed in July 1980, the major banks were overcome by the speed of change and were forced to compete with high interest rates offered by government bonds. They increased their interest rates to 40-50 per cent from the traditional 10 per cent. This move attracted savings back into bank deposits rather than into government bonds, property and other tangible objects of high investment value. It did not, however, stop the rate of borrowing, despite its high cost, because importers needed short term loans, and the industrial demand for loans persisted as industry struggled to keep going and to compete more effectively in the market place.

A direct result of this situation was the emergence of money brokers aiming to benefit from the high interest rates. The money brokers borrowed at high interest rates from the public, and lent the money to businesses which could not obtain bank loans, at much higher rates. The Turkish public responded rapidly to this new speculative spirit, but because of the highly perilous nature of this business, hundreds of thousands of people lost their savings when money brokers went bankrupt. Bankruptcies occurred because brokers could recoup neither the money nor the interest due to them, from businesses which had either gone bankrupt themselves or were not able to pay the interest when it was due. The collapse of Kastelli, the biggest money broking firm, was such a serious affair that it contributed to the removal of Ozal in the Summer of 1982 from his all powerful position as the economic supremo. Ozal was popular in the centres of international capital, and was chosen by *Euromoney* as the most successful economics minister of 1981, but pressures were building up, and the Kastelli crash was taken as a symptom of the unnecessary rigidity of monetarism. [2]

The demise of the largest private money broking firm on a wave of bankruptcies shook Turkish financial circles and precipitated a swing back to more stable and safer methods of transaction in money markets, and the new banking law of September 1982 introduced measures to safeguard regular banking procedures. Foreign banks were also encouraged to open branches in Turkey; Citibank, American Express, etc. These moves started as being seen as the beginning of a new phase in Turkish banking, as Turkish banks were gradually put into a situation where they had to come to terms with the fact that they had to compete not only with each other but also with big international banks. Merchant banking is an area where Turkish financiers have hitherto not made any inroads. This is certainly a potential growth area where the new law, which prevents the control of new banks by big holding companies, should help. [3]

In addition to forcing Turkish banks to become more competitive and efficient, foreign banks were also seen as likely to play a crucial role in export financing. Because of the long tradition of working in a closed economy and operating as parts of large industrial/financial holdings, Turkish banks had not had sufficient opportunities to gain experience in international money markets and were not used to the new liberalised trade regime. Foreign banks could fulfill a vital function by providing the expertise and facilities to integrate Turkish capital more fully with international capital. There are some Turkish banks which do possess the necessary expertise to operate in international money markets but they are still few and far between. Unless the injection of "new blood", proficient in foreign languages and aware of the needs of the new economic order in Turkey, can be realised successfully and fairly quickly, the old guard will find it very difficult to cope with the new trade regime and export financing, and international banks will have further reason to delay their entry into the Turkish market. This is one area where no real progress has been made so far.

Another area of development in the financial sector has been the entry of international financial consultants and accountants (e.g., Price Waterhouse, Arthur Andersen, Coopers & Lybrand, Peat Marwick, Ernst & Whinney, Arthur Young) into the Turkish market. This development is particularly significant for those Turkish firms which want to operate in world markets and also for foreign firms wanting to invest in Turkey or enter into joint-venture programmes with Turkish companies. Any large company which operates in Turkey from now on or which wishes to raise finance in international money markets will be influenced by the standards and practices laid down by these international firms. These firms will, therefore, influence the development of Turkish capital and its reproduction on a world scale, and can be seen as representing the "new face" of Turkish capitalism.

The export sector: Ozal, before he was deposed in 1982, had said that Turkish businessmen "will either export or die". In a society imbued with the ideology of a closed society and economy, this message was difficult to comprehend. Foreign trade accounted for a very small percentage of the GDP, and imports constituted a far greater proportion of the foreign trade than the exports (Table 1).

TABLE 1

Share of Exports & Imports In Total Volume of Trade

	Total Volume of Trade	Share of Imports %	Share of Exports %
1965	1036	55.22	44.78
1966	1209	59.42	40.58
1967	1207	56.72	43.28
1968	1260	60.60	39.40
1969	1338	59.88	40.12
1970	1536	61.69	38.31
1971	1847	63.38	36.62
1972	2448	63.84	36.16
1973	3402	61.32	38.68
1974	5310	71.14	28.56
1975	6140	77.18	22.82
1976	7089	72.35	27.65
1977	7549	76.78	23.22
1978	6887	69.78	30.85
1979	7331	69.15	30.85
1980	10577	72.50	27.50
1981	13636	65.51	34.49
1982	14480	60.32	39.68

Source: TUSIAD, 1983, p.81

The main cause of the sustained deficit in the balance of payments was the political decision to overvalue the Turkish Lira. This made imports for the manufacturing sector relatively cheap, in line with the policy of import-substitution. Exports were primarily of agricultural produce with some manufactured goods. With the drastic devaluation of the Turkish Lira (TL), Ozal broke the stranglehold of the import-substitution economy and forced businesses to become competitive in export markets or risk bankruptcy. As a result there was a substantial boom in exports with the manufacturing sector increasing its share (Table 2).

Ozal, and his successor Kafaoglu, saw the exchange rate as the crucial factor in promoting exports. But by making the TL a convertible currency and leaving it to float in international money markets, they began to play a dangerous game. If the gamble succeeds, however, the rewards could be quite substantial.

The impact of devaluation was soon felt by Turkish industry. Competition in the domestic market increased fiercely, while exporting remained difficult where the quality and cost of Turkish goods made them uncompetitive. Cash flow problems caused the dwindling of fixed and capital investment, and retarded the modernisation of production. Although strikes were now illegal, wages became increasingly burdensome where stock could not be sold, and many businesses were forced to close, giving rise to further unemployment.

The first major casualty was Guney Sanayi, one of Turkey's largest textile companies. When Ozal refused to rescue it with state funds he attracted very strong criticism from industrialists, and antagonised the bureaucracy which was accustomed to etatism. When the Kastelli collapse followed, the criticisms of Ozal became too widespread for the military to ignore. Ozal was replaced by A.B. Kafaoglu in July 1982. Kafaoglu was more of an accepted establishment figure and more in tune with the mentality of the military chiefs who were after all the products of an etatist state despite their attempts to break away from the constraints imposed by such a state. He was also perhaps more diplomatic, but it seemed unlikely that he would be able to change the implementation of the stabilisation programme, and enjoy the same degree of confidence shown to Ozal by various agencies of international capital. At the time when Kafaoglu took over the helm of Turkish economy, Turkey's export performance was still less than satisfactory compared to countries such as Mexico, Brazil, Portugal, Greece and Spain (Table 3 below).

The main markets for Turkish products were, and still are, at present in Western Europe, although exports to COMECON markets are growing. The main potential for expanding markets in future, particularly for manufactured goods, probably lies in the Arab Middle East, and in Asia. Turkey still has the advantage of being self-sufficient in food, despite the widespread inefficiency in the agricultural sector and, although government and industry are struggling at the moment to develop new markets, the potential of the manufacturing base is considerable. Once the stabilisation programme has been implemented, and its measures generally accepted, growth seems inevitable. Some trends are already evident: Turkish companies, for example, are some of the largest building contractors in, for example, Libya, Saudi Arabia and Iraq. Turkey has one of the largest fleet of TIR trucks in Europe, and these trucks are increasingly coming out of mothballs and back into service. The textile industry is so well established that, despite the EEC agreements, European countries had to impose tariff barriers to protect their own textile industries. Today, Turkish textiles are finding ready markets in the Middle East. The automobile industry is dependent on licensing agreements with some of the big European motor companies, and has an enormous share in the export market for manufactured goods, although it is badly in need of improved efficiency.

In order to promote the export of agricultural produce, support prices for agriculture are being increased more slowly than the rate of inflation, thus preventing agricultural producers from relying on state

TABLE 2

Sectoral Rates of Increase in Exports (%)

	1963/ 1967	1967/ 1972	1972/ 1977	1978	1979	1980
I. Agriculture & Livestock	9.9	7.3	11.4	48.1	-12.9	24.4
1) Cereals & Pulses	-4.2	39.8	27.2	117.6	-37.5	10.3
2) Fruits & Vegetable	11.0	8.3	17.4	27.4	15.6	16.4
a) Hazelnuts	9.0	6.8	16.6	31.8	6.7	11.9
b) Raisins	8.1	6.1	19.7	32.9	15.2	13.5
c) Others	11.8	14.4	17.7	14.0		27.2
3) Industrial Production	13.8	5.7	4.9	45.8	-29.2	37.6
a) Tobacco	15.3	2.1	6.1	28.1	-21.4	32.1
b) Cotton	13.4	7.8	1.9	65.8	-34.6	41.6
c) Others	-	18.1	22.2	2.0	- 4.4	40.0
II. Mining & Quarrying Products	17.6	11.1	29.1	- 1.4	6.5	44.2
III. Industrial Products	3.7	26.5	19.3	6.1	26.2	33.4
1) Food & Beverages	2.9	12.9	7.9	-25.4	37.3	38.6
2) Textiles	-	57.6	37.1	18.9	22.3	12.4
3) Forestry Products	-1.8	27.7	- 9.4	10.7	50.0	165.7
4) Hides & Leather	28.0	162.8	19.3	-22.8	10.0	13.5
5) Chemical Industry	15.5	32.4	26.2	-29.1	0	218.7
6) Petroleum Products	56.5	133.3	-100.0	-	-	-
7) Cement Industry	-100.0	+	- 9.6	330.9	2.3	-11.8
8) Glass & Ceramics	- 9.3	65.8	49.4	9.7	23.3	- 3.1
9) Non-Ferrous Metals	29.3	-18.7	27.7	-42.4	25.0	25.8
10) Iron & Steel	47.2	34.4	14.4	47.2	47.6	9.2
11) Metal Prod & Mach.	50.5	84.7	28.0	28.1	0	64.4
12) Elect. Appliances	+	74.2	28.0	22.0	0	154.2
13) Vehicles	24.5	86.7	102.2	-33.9	350.0	89.0
14) Others	19.8	40.4	5.3	9.5	100.0	43.7

Source: TUSIAD, 1981, p.125

TABLE 3

Export Performance of Selected Semi-Industrialised Countries

	Turkey	Mexico	Brazil	Portugal	Greece	Spain
Population (millions) 1978	43.1	65.4	119.5	9.8	9.4	37.1
Area (thousand km^2)	781.0	1973.0	8512.0	92.0	132.0	505.0
Per Capita GNP	1200.0	1290.0	1570.0	1990.0	3250.0	3470.0
Exports (million $) 1978	2288.0	5739.0	12527.0	2393.0	3341.0	13115.0
Annual Growth Rate of Exports 1960-70	1.6	3.3	5.0	9.6	10.7	11.6
Annual Growth Rate of Exports 1970-80	2.5	5.2	6.0	-5.9	13.1	11.0
Exports as % of GDP, 1960	3.0	10.0	5.0	17.0	9.0	10.0
Exports as % of GDP, 1978	6.0	11.0	7.0	20.0	17.0	16.0
Exports Fuels, Minerals % of Total, 1960	8.0	24.0	8.0	8.0	9.0	21.0
Exports of Fuels, Minerals % Total, 1977	8.0	32.0	10.0	4.0	14.0	6.0
Exports of Primary Comm. % of Total, 1960	89.0	64.0	89.0	37.0	81.0	57.0
Exports of Primary Comm. % of Total, 1977	67.0	39.0	64.0	26.0	36.0	23.0
Exports of Manufact. % of Total, 1960	3.0	12.0	3.0	55.0	10.0	22.0
Exports of Manufact. % of Total, 1977	25.0	29.0	26.0	70.0	50.0	71.0
Manufacturing as % GDP, 1960	13.0	23.0	26.0	29.0	16.0	27.0
Manufacturing as % GDP, 1978	18.0	28.0	28.0	36.0	19.0	30.0

Source: _TUSIAD_, 1981, p.129

subsidies and flooding the domestic market. These producers are being forced to look for new markets, which they have not had to do before. The danger with these policies is that the pressure to sell in new markets has come before any change in the experience and expertise of agricultural producers. The move from supplying Turkey's towns and cities to competing in Middle Eastern and European markets needs not only a change in outlook, but also changes in the organisation of the agricultural sector. Small producers are likely to take the simpler options of competing more fiercely in local markets, or abandoning agriculture altogether. Turkey cannot be developed as an export-oriented economy, however, without transformation of the agricultural sector, and I shall return to this problem below.

Foreign investment: Sustained economic growth cannot be achieved simply by persuading producers to export their goods, it is also necessary to attract foreign investment. Since the establishment of the Republic, nationalist sentiment and the closed economy had discouraged foreign investment in Turkey. Although a bill was passed in 1954, "Law 6224", to promote foreign investment in Turkey, it was never implemented. "Law 6224" was a victim of the etatist policies of the Turkish state and its bureaucracy, which persisted in obstructing foreign investment in spite of the existence of a legal framework for effecting changes.

TABLE 4

Sectoral Breakdown of Joint Ventures Operating in Turkey Under "Law 6224"

	No of Firms	Million TL Authorised Foreign Investment	Total Share Capital	Percentage of Foreign Capital in the Total
MANUFACTURING INDUSTRY				
- Food, Drink & Tobacco	9	1,068.75	1,937.13	55.70
- Textiles	1	374.10	499.22	74.93
- Paper	1	39.40	70.34	56.01
- Tyres	3	462.90	831.00	55.70
- Plastics	1	3.40	8.74	38.90
- Chemicals	20	992.13	1,343.68	73.83
- Glass	2	83.65	700.00	11.95
- Motor Vehicles	7	1,027.31	3,165.40	34.25
- Metal Goods	8	509.82	2,543.99	20.04
- Machinery Man.	8	1,135.08	2,853.75	39.77
- Farm Machinery	1	17.50	70.00	25.00
- Electrical Machinery	20	2,538.09	6,434.27	39.44
- Cement	2	174.00	580.00	30.00
- Construction Materials	1	0.60	30.00	2.00
- Shipbuilding	1	2.00	20.00	10.00
- Steel	1	100.00	1,000.00	10.00
- Forestry Prod.	1	150.00	500.00	30.00
AGRICULTURE	1	1.02	2.00	51.00
MINING	1	20.00	20.00	100.00
SERVICES				
- Tourism	7	358.51	863.85	41.50
- Banking	6	1,403.50	3,500.00	40.21
- Engineering & Consultancy	2	13.16	34.00	38.70

Source: TUSIAD, 1981, p.147

The protective trade regime and regulations restricting the transfer of profits abroad (despite "Law 6224") ensured that the little foreign investment that was attracted came in joint ventures and on a limited scale (Table 4 above).

From the start of the 1980 stabilisation programme, there was a marked increase in support for and encouragement of foreign investment. "Law 6224" is still in effect, and with the political stability provided by the military regime, it began to be complemented by other measures aimed at making its provisions effective. The Turkish Lira was floated, and a new banking law introduced. Red tape was drastically cut, and bureaucratic responsibilities withdrawn from the Ministries of Finance, Industry and Commerce, and the State Planning Organisation, and transferred to a new Foreign Investment Department. This department is attached to the State Planning Organisation, but has its own staff of highly paid specialists.

These measures were successful in attracting investment, and $300 million of foreign capital has been invested by 1982, although this was still far short of the scale of investment needed to stimulate sustained growth in the economy. One area in which the government made substantial changes in order to promote foreign investment was tourism. Under the new economic regime, foreign investment of up to 100 per cent was encouraged, with complete control of the project left to the investor. Efforts were also made to attract foreign investment into agriculture.

In spite of these efforts and their initial success, the measures taken are seen as inadequate by a number of organisations, including the World Bank and the IMF, as well as TUSIAD. In their 1981 report, TUSIAD propose new measures such as the introduction of a capital market law to reorganize the share and bond markets, and the introduction of free trade zones to take advantage of Turkey's geographical proximity to the Middle East. The report goes on to suggest that Turkey must learn to make good use of foreign capital investments, and choose those which will stimulate,

> "...progressive increases in local content and the most rapid possible passage from the assembly to the manufacturing stages; integrated schemes wherever possible - Basically export-oriented projects, which will help to increase Turkey's foreign currency earnings."
> (TUSIAD, 1981, p.150)

The report also suggests that every effort must be made to ensure that investments are realised on time, and that the agreed terms of investment are not changed.

The hoped for tourist boom is still to come, and foreign investment has not yet become the flood that is needed, but the fact remains that fundamental attitudes towards the desirability of foreign investment are changing, and this has already had some effect. Future success in attracting new foreign investment will depend not only on attitudes, but also on the effective implementation of practical measures to attract capital.

The State Economic Enterprises: Among the measures taken by the military regime, those affecting the state economic enterprises (SEE) generated the greatest controversy. These enterprises were the keystone of etatist policies, and the demise of these enterprises would not only signal the final demise of etatism as an economic strategy, but would also undermine the powerful position of the bureaucracy which flourished on etatist policies, and which has always obstructed alternatives. The effective use of state power depends on control of the bureaucracy, since without bureaucratic support policies will not be properly implemented (as the fate of "Law 6224" has illustrated).

The state economic enterprises own 47 per cent of Turkey's industry, and they have monopolies of cigarettes and tobacco, spirits and petrochemicals. They were originally set up to pioneer the development of Turkish industry during the early years of the republic, until private capital was sufficiently powerful and developed to take over all sectors of the economy (Walstedt, 1980). In time, bureaucratic control of these enterprises and the use of state subsidies to cover inefficiency and other shortcomings, turned them into permanent and inefficient features of the Turkish economy. Apart from blocking any moves made by private capital to enter the markets they dominated, they also provided a haven for nepotism, and were used by successive governments to repay political debts. When governments changed, the heads of these enterprises went with them. Incompetence was acceptable, but political disloyalty was not.

Any major shift in the development of Turkish capitalism must require, therefore, a diminishing role for the SEEs in the economy as a whole. Because of the political nature of the enterprises, this will not be easy to achieve, but it became generally recognised after the 1980 coup, that the activities of these enterprises will have to be taken over by private capital in the long run. In the meantime, the new measures of the stabilisation programme forced the state enterprises to operate under market conditions without any state subsidies. The eventual reduction of state participation in the areas of economic activity covered by these enterprises will dramatically alter the sectoral organisation of the economy, particularly in the manufacturing and service sectors. [4]

Taxation: Any move to rationalise the economy and to achieve sustained growth in Turkey, must include reforms aimed at restructuring tax thresholds. Wage earners and groups on fixed incomes, who have their income tax deducted at source, paid two thirds of the total revenue, while their share of the national income was only 30 per cent. Agriculture and business contributed very little revenue and, because of the political strength of these interests, there was very little that the state could do to remedy the situation. After 1980, measures were taken to reform the tax structure, and to make it more equitable, as inflation was exacerbating existing inequalities.

In the stabilisation programme, the basic rate of income tax was cut to 40 per cent, on all incomes up to TL2 million *per annum*, with a further 1 per cent each year for the next five years. Major revisions in taxation were planned for 1983. The main changes were intended to be in the agricultural sector, which will now be heavily taxed.

Hitherto, farmers could claim a minimum 70 per cent of their gross income as expenses and absentee landlords could also do so as if they were actively engaged in farming. Now 70 per cent is the maximum rate for expenses claims. Absentee landlords will not be able to claim any expenses at all, instead they will be taxed on the basis of their rental income. In addition, a withholding tax of 5 per cent will have to be paid an all sales made by farmers either to merchants or in public market places. The law defines, "small farmers", who are exempt from tax, but farmers owning a tractor above 25 bhp or a harvester, or whose land exceeds a certain size or whose annual sales receipts exceed TL500.000 have to submit an income tax declaration.

The rate of corporation tax was to be raised to 50 per cent for capital companies and co-operatives, and for all others to 35 per cent. The new rate incorporated both the previous 25 per cent corporation tax and 20 per cent income tax.

The new law also covered other areas of taxation such as real estate purchase tax, property tax; inheritance and gift tax; vehicle and vehicle purchase tax; sales tax, production tax and value-added tax; taxes on banking and insurance transactions; capital gains tax on property, etc. The most important change, apart from altering the basis of income tax, was the introduction of value-added tax was to be modelled on the German system and was aimed at reducing tax evasion which had been widespread.

These changes are not really as drastic as one might expect them to be, given the scale and speed of change in other sectors of the economy. In addition, the bureaucracy is not geared to change even of this limited nature. There are bound to be delays and confusion in the assessment and collection of tax which will prevent these reforms being fully effective.

Agriculture: It was understood by everyone that the stabilisation programme could not achieve real changes in Turkish economy and society unless effective measures were also taken to transform the agricultural sector. Agriculture acts as a constraint on industrial development in a number of ways. The sector as a whole is relatively inefficient and weakly incorporated into the exchange economy, and State subsidies have made agricultural producers insensitive to market mechanisms. Agriculture no longer dominates the Turkish economy, but the agricultural sector still contains more than half the population and has been an important earner of the foreign exchange on which industry is dependent.

Production is still characterised by low productivity and feudal relations of production. Some landlords have emerged as capitalist farmers, or market gardeners, using capital-intensive technology, but relations of production do not necessarily change. A tenant who drives a tractor does not necessarily change his relations with his landlord. The big landlords have very largely retained their class power, and have managed to impede the processes of accumulation by monopoly/industrial capital.

Agriculture, apart from being the most inefficient sector of the economy, is also the sector where there is extreme inequality in the distribution of income. [5] The majority of the wealthy in Turkey are, by and large, still big landlords and big farmers who have always paid very little tax. At the same time extreme poverty and the lowest income groups are to be found in the agricultural sector. For many years, the landless, dispossessed and small peasants have been migrating to the cities, where some find work, but with growing unemployment many have come to subsist below the poverty line. These concentrations of disadvantaged population in shanty towns round the cities have led to severe social and political problems. These areas were frequently centres of resistance to state authority prior to the military intervention in 1980.

Attempts to improve efficiency in agriculture cannot leave out of account the extremes of income inequality and their political consequences. Any policy aimed at reducing inequalities in the agricultural sector would have to confront the overall inequality of income distribution which has characterised Turkey for generations.

In the agricultural sector, feudal relations of production will have to be dismantled and small direct producers will either have to have their own land or be transformed into wage earners. Landlords will become merely landowners, and unused land be reallocated to the landless, thus creating the basis for a politically acquiescent rural class. The development of opportunities for wage earning and land ownership in the rural areas should stem the flow of emigrants and even attract some population back from the shanty towns.

The transformation of the agricultural sector is not simply a question of improving efficiency and reducing income inequality; the place of agriculture in the Turkish economy has to be changed. If agriculture is to promote rather than inhibit the expansion of Turkish capitalism, the agricultural sector has to be fully integrated with and subordinated to the market mechanisms of the exchange economy. Agriculture will have to dominated by monopoly/industrial and financial capital. These changes must have an extensive impact on the structures of Turkish economy, society and culture.

The military regime embarked on preliminary changes in the agricultural sector, using taxation and land reform. The measures taken to levy taxes on agriculture have been discussed above, so I now turn to the problems of land reform.

The aim of land reform is to allow the most efficient exploitation of available land, with some redistribution of resources. Previous attempts at land reform in Turkey foundered on the political strength of the landed interests. A pilot project was established in Urfa, Southeast Turkey, in early 1970s but was unsuccessful in achieving its aims, as the pilot project was supposed to create a set of policies which could be spread to other areas. Urfa was chosen because the capitalisation of agriculture was scarcely begun, and feudal relations of production were dominant, but it was precisely these characteristics which limited its success.

In other parts of Turkey, agrarian reform and the mechanisation of agriculture raised productivity but without any redistribution of resources. This led to the creation of a stratum of landless peasants, which widened rural inequalities, and drove a large proportion of the population into the cities or abroad. The landed interests could not prevent the passage of land reform legislation through parliament, but they could prevent the law from being implemented effectively. The military set out to revive land reform policies in Urfa and elsewhere, and as a first step had directly to confront the landed interests, which would have been unthinkable before 1980.

There is no way of knowing at this stage whether land reform accompanied by the reform of agrarian structures will succeed or will become severely restricted. At present, censorship of information from Eastern and Southeastern parts of Turkey makes it very difficult to assess progress, but it seems that reforms were soon being ruthlessly and efficiently promoted. Successful reform will bring widespread social change, and it is unlikely to proceed without considerable obstruction and social upheaval, and much depends on the eventual outcome.

One other consideration which ought to be mentioned is that since the early 1950s the military itself has had growing commercial and industrial interests. It is safe to assume, therefore, that the military will benefit from the transformation of Turkish capitalism if and when it is effectively completed. Today, the military's industrial and commercial holdings are some of the largest enterprises in the country. Their future growth depends, as other privately owned large industrial and commercial holdings' future do, on the successful outcome of the present stabilisation programme. The transformation of agriculture is obviously an urgent priority to be realised to this end.

Political measures

The package of economic measures accepted by Demirel in 1980, formed the basis for the the military regime's economic strategy, but these measures could not successfully be implemented without corresponding political changes. Turkish democracy failed when it was unable to provide an appropriate political framework within which economic changes could be realised, and the military were faced with the task of creating an alternative political framework which would permit the opening of the economy. As a military dictatorship with no scruples about the abolition of civil liberties, the regime did have the power to instigate the necessary political changes. The military has made it clear that it is committed to the preservation of the undisputed dominance of monopoly/ industrial and financial capital at the political level. This situation has been achieved for the first time in the history of Turkish capitalism, and there is no sign that the new civilian government led by Ozal, will want to change this situation. The persistent contradiction of the non-correspondence between the economic power of the monopoly/industrial and financial capital and their political power has finally been resolved, but only at the cost of the democratic system. [6]

The democratic basis of the parliamentary system which existed before the 1980 coup was safeguarded by the 1961 constitution. This constitution gave the commercial and agricultural fractions of capital to have access to state power, thus preventing the dominant fractions of capital, monopoly/industrial and financial capital, from monopolising state power. This situation could only be resolved by restructuring the state in accordance with the restructuring of the economy, thus allowing the dominant interests to monopolise state power. In order to maintain this new coincidence of economic and political power, the military regime had to construct a new constitutional framework for Turkish society. The 1961 constitution created the conditions for a struggle over state power, and the regime tried to replace this with a framework for stability, in which state power is securely monopolised by the dominant fractions of the bourgeoisie, backed by the military.

The new 1982 constitution was presented as a new step in a series of experiments in Turkish democracy, but the regime had no intention of leaving this experiment wholly in the hands of the people. In October 1981, a Consultative Assembly, with members handpicked by the military, was given the task of drawing up a new constitution in which the military would be recognised as an effective force in the civilian state. Political opposition to this new power structure was plainly not to be tolerated. Five hundred and ninety four former politicians and others were banned from taking any part in politics, including, Ecevit, Demirel and Turkes. [7] Ecevit was the only one of the old politicians who had voiced any criticism of the regime, and he was gaoled twice with further cases against him pending. Turkes and his accomplices were put on trial with the public prosecutor demanding the death sentence for a number of defendants,including Turkes. This was not very surprising, as the public prosecutor generally demands the death sentence and Turkes and the Nationalist Action Party (NAP), having fulfilled their role admirably, were of no further use to the ruling class, and were indeed becoming something of an embarrassment. The fact that the NAP had no role to play in the present political experiment in Turkey does not mean that the threat of fascism is dead. Fascism has changed its form, but it could be revived at any time if it were needed.

All the political parties which had existed before September 1980, were banned and stripped of their assets. Political discussion and criticism of the new regime was made illegal although, by observing strict rules stipulated by the government, select groups and individuals could express opinions. There was effective censorship of the mass media, people seen reading illegal material were liable to be arrested, and there was no right of demonstration. Open criticism was effectively silenced with the detention of dissidents, trade unionists, politicians, and even liberal intellectuals and defence lawyers.

Immediately after the coup, all trade union activity was banned, and all strikes made illegal. To pacify the workers, who were suffering the effects of very rapid inflation, a flat rate wage increase of 70 per cent was granted for all wage contracts then under negotiation. The progressive confederation of trade unions, DISK, was abolished and its leaders put on trial, with the prosecution demanding the death sentence. [8] The alternative confederation of trade unions, Turk-Is, which was closely identified with business interests, was not abolished, but was banned from taking part in trade union activity. Turk-Is was left as the only voice of the trade union movement, and the deep divisions within Turkish society were papered over, at least temporarily, by not allowing their existence to be publicly acknowledged.

THE NEW POLITICAL FRAMEWORK

As members of the Consultative Assembly went about their task, it quickly became clear that the new constitution was going to be very different from the 1961 constitution. The 1961 constitution was drawn up with the aim of maintaining a balance between the executive, legislative and judicial functions of the state, while the 1982 constitution was clearly intended to give supremacy to the executive. The overall objectives were efficiency and expediency. Policy-making and implementation was to be a rapid process which could be used to advance the transformation of Turkish capitalism. By reducing the legislative functions of the state to a secondary role, the new constitution was to reduce not only the time taken to enact new legislation, but also any "unnecessary" political conflicts which might arise in the process. The judiciary was to be restructured to complement the reorganisation of the state apparatuses; the execution of state power was to be located in the Office of the President, thus giving the President enormous direct powers which the 1961 constitution had deliberately avoided; relations between the citizen and the state were to be restructured with the wishes of the citizen clearly subordinated to the needs of the state. The safeguards built into the 1961 constitution, which gave citizens some means of controlling state power through legislative and judicial processes, were swept away.

The military saw themselves as the Kemalist guardians of the new socio-economic and political order, and they must accept full responsibility for the problems which this new constitution will provoke. The 1961 constitution created problems in a society where the capitalist classes were not yet crystallised, and its democratic framework could not contain the conflicts of interest which developed, but the 1982 constitution, in trying to correct these shortcomings, contains the seeds of far more destructive conflict. The new constitution deliberately and systematically excludes the subordinate classes from any effective participation in the political decision-making processes. Once power is restricted in this way, any attempt to broaden power sharing is likely to unleash very destructive political conflict which could lead to the collapse of the present social order. In spite of the present general acceptance of the military intervention, and an apparent return to representative system of government in 1983, the reduction of civil rights remains, and can only be maintained in the long term by the use of force and repression.

It is scarcely surprising that when the provisions of the new constitution were revealed to the public in July 1982, they attracted immediate criticism. In spite of the nature of the constitution, it was presented as a new version of democracy, and some groups from the universities, the professions and business organisations, were allowed to comment publicly on the proposals. There was almost unanimous condemnation of the proposals, and their probable consequences. Even within the Consultative Assembly, bitter disputes developed over the proposals, but the views most sympathetic to the intentions of the military won out, and General Evren gave his personal approval to the new measures.

Evren's next step was to try to sell the constitutional proposals to the electorate as the only hope for Turkey's future. He toured the entire country, with extensive coverage of his tour by the media, driving home the message that the 1961 constitution had contributed to the degeneration of Turkish politics; ex-politicians were guilty of pushing the country to the brink of civil war, and that intervention by the military gave the nation one last chance of regaining its rightful position in the world. He presented the new constitutional proposals as the means of preventing the country from sinking to similar depths in future.

A referendum was held in November, 1982, and the Turkish people were urged to vote "Yes" for the constitution and, by implication, to elect General Evren President of the Republic for a seven year term. Evren's ability to communicate ideas to the people effectively, and his personal appeal as the man of action who, like Ataturk before him, had saved the country from collapse, go some way towards accounting for the overwhelming support given to the constitution in the referendum, when 91.2 per cent of the electorate voted "Yes". The referendum incorporated a number of procedures designed to ensure a large affirmative vote, such as different coloured papers for "Yes" and "No" votes, and the threat of legal action against anyone who did not vote, but the support given to the constitution and to Evren personally cannot be explained away simply as an instance of electoral manipulation. Voting must have been influenced by the general fear of economic chaos and the possibility of renewed terror, and any alternative viewpoints were silenced. There were no demonstrations against the referendum as all such activity had been banned by the military, and the military also made very effective use of the mass media to the exclusion of other voices.

Ever since the wide acceptance of the new constitution by the Turkish people, the military regime has been active in establishing its control over the civilian apparatuses of the Turkish state. The President, General Evren, has wide sweeping executive powers which gives him the ultimate power of veto over anything that may be seen as not conducive to the maintenance of the new regime. In order to increase the legitimacy and the effectiveness of the state in implementing, co-ordinating and monitoring state policies, Turkey has been divided into eight new regions, each with its own governor. These regional governors have been given very wide powers, and they control the sixty-seven provinces of the Turkish administrative and political system between them. The provinces are also governed by a provincial governor, and these governors have also been entrusted with unprecedented powers to be used in cases of civil strife and political unrest. The regional governors are mostly high ranking, retired military officers or are handpicked by the military for their political reliability.

The political situation today is potentially highly unstable. Any serious attempt to move towards a political system resembling a democracy has been pre-empted by the actions of the military regime. The military have been totally insensitive to the growing uneasiness which is evident among the Turkish people, as an increasing proportion of the public have slowly but steadily started questioning the existing order. This was shown in the widespread refusal to support the approved party in the 1983 election. The military misinterpreted the support which they received in the November 1982 referendum when an overwhelming majority had voted "Yes" for the new constitution, and underestimated the desire for a return to stable civilian rule.

Turgut Sunalp, a retired army general, was picked by the military to form a new political party, the Nationalist Democracy Party. This party had the blessing of the military in forming the first civilian government, and was personally promoted by Evren. In addition to actually forming a party of their own, the military also did everything possible to prevent the formation of other political parties. Hundreds of candidates for the new parliament who were members of the newly created political parties were prevented from standing. The military even went further and closed down a number of parties, a few weeks after they had been legally formed, without feeling any need to justify their actions. One of these parties, the Right Way Party, represented an attempt by the old Justice Party politicians to participate in the new political process, and had Demirel's support. When the party was closed down, Demirel himself was taken into custody for violating the law banning the politicians of the pre-1980 era from taking part in politics. It was made very clear that civilian rule could only be allowed within the framework specified by the military.

The military took additional measures to preserve the political purity of the new era. They prevented Social Democratic Party (SODEP), founded by former members of the Republican People's Party from taking part in the 1983 election, but allowed the formation of two other political parties which they considered to be politically safe. These were the Motherland Party, led by the deposed economic supremo Turgut Ozal, and the Populist Party (PP), led by Necdet Calp who was a high ranking bureaucrat in the prime minister's office after the 1980 coup. Neither of these parties were significant additions to what the Nationalist Democracy Party had to offer, in terms of providing a wider spectrum of political policies, and they posed no threat to the military's definition of democracy. There have been also attempts at merger of SODEP and PP to enhance their political fortunes, but they have ended in failure.

As the election day of 6th November, 1983 approached it became increasingly clear that the military's choice was being rejected by the people. Television debates between Sunalp, Ozal and Calp led to a further decline in the fortunes of the retired general. Public opinion polls put Ozal and his Motherland party in front, with the Populist Party taking the second place and the Nationalist Democracy Party (NDP) trailing in third place. President Evren took, by all accounts, high-handed measures in, firstly implying that Ozal was a liar, and secondly offering Sunalp to the people as the only alternative. The military also banned the public opinion polls as they were confirming a steadily widening gap between the Motherland Party (MP) and the other two. On the eve of the election, Evren went on television to broadcast to the Turkish people and to remind them of the military's choice and to denounce Ozal once again. This seemingly desperate act was probably counterproductive, and Ozal came to power with an absolute majority (MP-184 seats, PP-92 seats, NDP-54 seats). [10] Ozal had won the election after the humiliation of having been sacked by the military, but his triumph was cautious. He was faced with the problem of having to operate under the close supervision of President Evren, with precious little room to manoeuvre on his own.

The Motherland Party had the backing of a number of influential centres of international capital, e.g. Wall Street, and the EEC as well as of some apparently contradictory allies, e.g. Libya. After the results of the elections were known, the IMF and other international organisations expressed their relief and their confidence in Turkey and its future progress. Foreign capital investments were increased and the IMF sent a new list of demands, primarily asking Turkey to increase its exports by 10 per cent, to adjust the value of Turkish Lira to a realistic level, i.e. to devalue it, to relieve the Treasury of the burden of the state economic enterprises, to increase the rate of interest to 40-45 per cent and not to exceed the monetary limits imposed by the Central Bank. [11]

After the election the general level of prices shot up causing great hardship, and unemployment increased to an official figure of 20 per cent. The rate of inflation went into an upward spiral. The policies that Ozal had pursued before he was sacked and replaced by Kafaoglu, formed the basis of the MP government's economic strategy. Ozal brought back the same people to manage the economy, and undiluted "monetarism" became established as the road for the further development of Turkish capitalism.

Ozal found himself in a difficult position because the major state apparatuses had already been put under the control of the military and the President retained the right of vetoing any government policies. The majority of the public do not expect this civilian rule to survive, and if this is the case, no political regime can reasonably expect to maintain its legitimacy and effectiveness for any length of time, within the present framework.

The response of the military to the challenge of establishing a democratic regime was rather heavy handed. The generals have discovered to their chagrin that the public prefer not to be treated like privates. The uneasiness that the military had felt about the transfer of their control into civilian hands became increasingly evident as the election date drew closer. Evren made it very clear that the military did not want a coalition government, and once the elections were over, one of the first things that the military did was to extend martial law for another period of six months and introduce a law extending press censorship. [12]

With heavy censorship and nationwide martial law still in force, it is virtually impossible to analyse the intentions of the military systematically. All the signs, however, point to the fact that the balance of power within the ruling National Security Council has swung in favour of the reactionary clique. This has made the direction of future developments perhaps more predictable, as the use of force against civilians is likely to become more frequent.

Since the critical issue of deciding when force is necessary always determines the legitimacy of a political regime, this new experiment in Turkish "democracy" is no exception to the rule. In circumstances like these any civilian government is likely to find working under an umbrella of military approval very difficult. This situation is likely to continue until monopoly/industrial and financial capital finalise their monopolistic control over the use of state apparatuses. Forming a government is not the same as controlling the state, and different governments have different degrees of access to and control of state apparatuses. Since the military is part of the state, it is yet another state apparatus and always directly involved in the maintenance of social and political order. Where the uneven development of capitalism gives rise to competition within the ruling classes for the control of the state, the military will have a much more open part to play than is the case where one alliance of class fractions has gained control.

Since the creation of the Republic, the Turkish military has taken on the role of guardian of public order, and it has its roots in Ataturk's _etatist_ and closed society. While the military has now facilitated a bold programme of economic and social transformation, it has done so very much in the spirit of its traditional role, and sees these changes as necessary for the maintenance of social and political order. It is not yet at all clear that the traditional organisation and _Kemalist_ ideology of the military will prove sufficiently flexible to allow the rapid expansion of an outward-looking economy.

CONCLUSION

In this chapter, I have looked at the consequences of the 1980 coup in the light of the intention of the military regime to effect a rapid transformation of the Turkish economy from inward-looking to outward-looking, within a new institutional framework which could create and maintain political stability. In pursuing these aims the military have brought about a new and remarkable correspondence between the political and economic power of the dominant fractions of the bourgeoisie, which has abruptly ended the long and destabilising struggle for the monopolistic control of state power. The military have been extremely consistent in implementing measures to further these aims, but it is still too early to tell whether or not their measures will achieve the desired effects in the long run. The transformation of an economy requires a series of simultaneous changes on many levels of economy, society and polity, and there are clearly a number of obstacles to change which have not yet been dealt with effectively, such as the inertia of the bureaucracy, the inefficiency of the state economic enterprises and the political power of the landed interests.

In addition to the many problems of achieving internal changes, the military regime is also constrained by Turkey's position in the world capitalist system. Since the world system is not a Parsonian unity, the centres of international capitalism generate divided interests, and Turkey may well be caught between the need to satisfy different interests emanating from, for example, the EEC, the United States of America, and a deeply divided Middle East. When Ozal was dismissed in the Summer of 1982, his team responsible for finance, state planning and the Central Bank also went with him, but his successor, Kafaoglu, did not seem to have a substitute team which could operate with comparable skill and certainty. Kafaoglu, after having worked very hard to reassure the IMF and other agencies of international capital and failed, had pursued a cautious policy of stabilisation, which committed the state to playing a central role. The present regime, led by Ozal under the auspices of the military, and its successors will remain dependent on the support of the international community, which has played such an important part in stimulating the move to an outward-looking economy. Various European governments and agencies have criticised the regime's stand on human rights, but this has not materially altered support from the centres of international capitalism enjoyed by the military and the Ozal government. It is not yet clear how long this level of support can be maintained, as the various internal obstacles to transformation are tackled, particularly if increasing force has to be used against the people.

The regime has had sufficient popular support since the coup to enable it to silence all opposition and to dismantle the democratic system, but the decision to prevent workers and peasants from forming their own political parties, and the blocking of any representation of the Turkish left in the new democracy represents a dangerously short sighted assessment of the problems of transforming Turkish capitalism. It is unwise for the ruling classes in a capitalist state to forget that the basic contradiction generated by the capitalist mode of production is between the interests of capital and the interests of labour. Turkey is no exception to this rule. Efforts to strengthen the reproduction of capitalism will intensify this contradiction and create an increasingly explosive social and political situation, as the contradiction becomes consistently apparent.

No capitalist state can resolve this basic contradiction and remain capitalist, but there are a variety of means of mystification, for example, by differentiating the interests of labour, by allowing labour to benefit from the the development of capitalism, and by reducing the visibility of the contradiction through various socio-economic measures, which would not challenge the class base of political power. The 1961 constitution, although liberal in spirit, did not lead to any effective participation or representation in politics by the subordinate classes. Ecevit was the only political leader who understood the explosive nature of this problem, and his attempts to extend the benefits of capitalism to workers and peasants were systematically frustrated. The 1982 constitution shows no attempt at all to recognise the interests of the working class, or of other subordinate classes and must, therefore, intensify the contradictions inherent in the capitalist mode of production still further.

The Turkish left is also characterised by a general confusion as to where the contradictions of capitalism are located. [13] The majority of the left have been operating with simplistic and mechanistic notions of underdevelopment in which either the existence of Turkish capitalism is denied and capitalist development is seen as occurring in the centres of capitalism at the expense of the development of the periphery or semi-periphery, of which Turkey is a part, or Turkey is seen as having a stunted form of peripheral capitalism created by the needs of imperialism. [14] In either case, imperialism is seen as the main cause of the lack of development in Turkey, and the international bourgeoisie is the main enemy. [15] The left, therefore, can only see the development of capitalism as being against the interests of the Turkish people, and by failing to recognise the strength of Turkish capitalism, adopts naive notions of how capitalism might be overthrown. It would be more realistic for the left to acknowledge how deeply rooted Turkish capitalism is, and also to see it as a potentially progressive force, for example, in destroying feudal relations of production and in generating the growing crystallisation of social classes.

Once capitalism is deeply rooted, and the state secure, urban guerrilla warfare and peasant uprisings become increasingly ineffective, and lead to the physical destruction or internment of the left. The left is more likely to survive as an effective force, and to be able to organise, by accepting the reality of capitalist development and the power of their own bourgeoisie. They can then struggle to operate legally within a representative political system. It would be better for the left to struggle for this end than to pursue apocalyptic dreams which they do not have the power to realise, and which so far have led only to their own destruction.

In the meantime, given the political conjunctures of Western capitalism, which Turkey is a part, this new Turkish democratic experiment is likely to continue as a facade concealing a system of very limited representation backed by force. The apparent acquiescence of the people should not lead us to assume that the divisions within Turkish society have vanished. The conditions which produced terror on the streets still exist, and there is always the danger of secessionist movements, particularly among the Kurds. In addition, the "Armenian problem" apparently receives considerable external support from Turkey's immediate neighbours. The agencies of the world capitalist system see their task as that of strengthening Turkey's economy, but they have not seen any alternative to accepting the internal contradictions which may be generated in the process, and this may ultimately prove to be disastrous.

The outcome of the enormous task of transforming Turkey from an inward-looking to an outward-looking economy must remain uncertain. Whether or not sufficient effective political leadership and managerial expertise actually exists in Turkey, the economic changes which have been initiated seem likely to continue, since no other option is available. The Turkish people have so far bowed to the inevitable, but whether the nation can support the social and the political stress of this transformation without major disruption remains to be seen. Events in Turkey are being closely watched by the international community, and it should not be forgotten that recent attempted coup in Spain was to have been modelled on the Turkish example if it had been successful. Spain is not the only country where the military are observing the progress of their Turkish counterparts.

NOTES

1. Turgut Ozal, before his removal from the position of economic supremo, said in an interview with Ertugrul Soysal that the country was undergoing the greatest crisis in its history. He blamed the economic policies of previous governments and boasted that the measures taken by the present government to resolve the crisis were more severe than the policies recommended by the IMF. He also suggested that conditions in 1982 would improve, but that 1983 would be the year when the balance of payments problem would be corrected, and investment and the unemployment situation would improve. Milliyet, 10th January, 1982.

2. At the same time the Turkish economy was named "The Most Improved Economy of 1981", Euromoney, October 1981.

3. Supplement to Euromoney, February 1982, p.21.

4. Bulutoglu, 1980, sees the maintenance and the reform of State Econ-

omic Enterprises as being vital for the development of Turkish economy in an inward-looking model, and the prevention of their privatisation as the ultimate goal. This view was part of the scenario put forward by the Ecevit government to fend off the IMF in 1979, and Bulutoglu, a professor of economics, was the Minister for Enterprises at the time.

5. A useful examination of income inequality in Turkey can be found in Dervis and Robinson, 1980.

6. This situation came to be generally accepted as unfortunate, but necessary, by international public opinion. In the U.K., this tendency was clearly reflected in two reports which appeared in the Financial Times (Monday, 18th May, 1981) and in the Guardian (Friday, 21st August, 1981).

7. Milliyet, Tuesday, 2nd November, 1982.

8. The plight of DISK repeatedly been taken up in the European Assembly by representatives of social democratic parties in Europe, and has been widely reported in the press of European trade union movements.

9. See the Appendix.

10. Milliyet, Monday, 7th November, 1983.

11. Milliyet, Monday, 21st November, 1983.

12. Milliyet, Wednesday, 9th November, 1983.

13. For a critical assessment of the Turkish left, see Samim, 1981.

14. It is not possible to review here all the relevant literature, but examples of the use of this framework can be found in Berberoglu, 1982; Pamuk, 1981, and Keyder, 1979 and 1981.

15. The contradiction between imperialism and the people of a social formation only becomes a basic contradiction under exceptional circumstances, such as, the invasion of Turkey by the Allied Powers after the First World War, the Chinese liberation struggle against Japanese imperialism, or the Vietnamese national liberation wars against the French and the United States imperialism. In these exceptional conjunctures, other internal contradictions are subsumed under this basic contradiction. But these other contradictions are, in addition, determinants of internal class struggles, which at the same time help to determine the nature of national liberation and independence movements. To interpret these historically specific conjunctures as typical processes of capitalist transformation is theoretically misleading and politically dangerous. The present situation in Turkey can never be understood simplistically in terms of American imperialism engineering coups, and underdeveloping the Turkish economy.

APPENDIX: MAIN PROVISIONS OF THE 1982 CONSTITUTION

The main provisions of the 1982 constitution can be grouped under five headings: the Presidency, the Legislature, the Executive, the Judiciary and Civil Rights.

I. THE PRESIDENCY

The Office of President is vested with the ultimate power for controlling the state.

1. The President is elected for a single seven year term of office, and does not have to come from the existing members of the House of Representatives.

2. The President will appoint and dismiss the Prime Minister.

3. The President will be assisted by an advisory council comprising former Presidents, Chiefs of Staffs, former Presidents of the Constitutional Court and members appointed by the President.

4. The President will appoint all members of the Constitutional Court, High Court Judges and Public Prosecutors, the Chief Public Prosecutor to the Court of Appeal, one quarter of the members of the Court of State, twenty members of the State Advisory Council, all the members of the State Inspectorate, the President of the Central Bank, the Director of Turkish Radio and Television and three members of the executive committee, and all initial members of the Turkish Academy of Language.

5. The President will have the right to dissolve the House of Representatives in the case of the Council of Ministers having been dissolved, and not being reconstituted within thirty days.

II. THE LEGISLATURE

The powers of the Legislature have been drastically curtailed, and the organisations of this branch of the state has been fundamentally changed.

1. There will be a single House of Representatives (*Meclis*) with 400 members (as opposed to the two tier system of *Meclis* and Senate which existed under the 1961 constitution.)

2. Elections will be held once every five years (as opposed to previously once every four years.)

3. The *Meclis* will have only a three months recess period.

4. The *Meclis* will need a quorum of one third of its members and will make decisions on the principle of simple majority.

5. People with less than eight years of formal education cannot become members of the *Meclis*.

6. Members of the Meclis who leave one political party will not be able to join another political party, nor will be eligible to become ministers.

III. THE EXECUTIVE

The executive is subordinated to the Office of the President and its structure is significantly reorganised.

1. The Prime Minister can call on the President to dismiss ministers.

2. An Economic and Social Council will be formed to advise the Council of Ministers.

3. The High Council of Arbitration, the State Inspectorate and the Turkish Academy of Language will become Constitutional organisations.

4. It will become the duty of the state to promote sport, maritime activities, flying and to protect environmental health.

IV. THE JUDICIARY

The function of the Judiciary is also changed to correspond with the increased powers of the President and the close links created between the Executive and the President.

1. The Court of Appeal will have ultimate power over the decisions of the Judiciary.

2. State Security Courts will be reinstituted.

3. Applications to the Constitutional Court for the annulment of executive action can be initiated only by the chairman of the opposition party supported by one fifth of the members of the Meclis.

4. Death sentences will be ratified directly by the President without having to be referred to the Meclis.

V. CIVIL RIGHTS

Changes in civil rights indicate a shift in power in favour of the state and against the citizen.

1. The rights and freedom of citizens are treated as inseparable from the integrity of the state and the need to maintain public and social order. These rights can, therefore, be curtailed when deemed necessary by the state.

2. Citizens, when deemed necessary by the state, can be asked to undertake work regardless of their own personal wishes.

3. For purposes of self defence or in circumstances affecting the defence of the state and public order, the taking of life will not be considered unconstitutional.

4. The state has the right to take any necessary action against civil associations, if their activities are deemed to be contrary to the interests of the state.

5. Education for citizens will be provided free of charge up to university level, but university education will not be free. The state will provide scholarships and loans to deserving students. The private sector will also be encouraged to take part in the education of the Turkish people. Education will be organised along lines laid down by the state according to its assessment of the needs of the national economy.

6. Members of associations and professional bodies cannot take part in demonstrations instigated by organisations other than their own.

7. Trade unions cannot take part in politics in any form and cannot pursue political needs.

8. Strikes and lockouts are not to be used to damage national wealth.

9. Political parties cannot be representatives of social classes, races or religious movements.

10. The voting age will be twenty-one.

REFERENCES

Berberoglu, Berch (1982); Turkey in Crisis, Zed Presss, London

Brett, Edward (1983); International Money and Capitalist Crisis, Heinemann/Westview, London and Boulder

Bulutoglu, Kenan (1980); Bunalim ve Cikis (Crisis and Escape), Tekin Yayinevi, Istanbul

Dervis, Kemal and Robinson, Sherman (1980); "The Structure of Income Inequality in Turkey (1950-1973)" in E. Ozbudun and A. Ulusan (eds.), The Political Economy of Income Distribution in Turkey, Holmes & Meier, New York

Euromoney (1981); October

Euromoney (1982); Supplement, "Turkey: Will the Experiment with Capitalism Work?", February

Financial Times (1981); Monday, 18th May, London

The Guardian (1981); Friday, 21st August, London

Keyder, Caglar (1979); "The Political Economy of Turkish Democracy", New Left Review, No.115, pp.3-44

Keyder, Caglar (1981); The Definition of a Peripheral Economy: Turkey, 1923-1929, Cambridge University Press, Cambridge

Milliyet (1982); 10th January, Istanbul

Milliyet (1982); Tuesday, 2nd November, Istanbul

Milliyet (1983); Monday, 7th November, Istanbul

Milliyet (1983); Monday, 21st November, Istanbul

Milliyet (1983); Wednesday, 9th November, Istanbul

Pamuk, Sevket (1981); "The Political Economy of Industrialisation in Turkey", MERIP Reports, No:93, pp.26-30

Samim, Ahmet (1981); "The Tragedy of the Turkish Left", New Left Review, No:126, pp.60-85

TUSIAD (1981); The Turkish Economy, TUSIAD, Istanbul

TUSIAD (1983); The Turkish Economy, TUSIAD, Istanbul

Walstedt, Bertil (1980); State Manufacturing Enterprise in a Mixed Economy: The Turkish Case, World Bank/Johns Hopkins University Press, Baltimore

Subject index